Inductive Links for Wireless Power Transfer

Pablo Pérez-Nicoli • Fernando Silveira
Maysam Ghovanloo

Inductive Links for Wireless Power Transfer

Fundamental Concepts for Designing
High-efficiency Wireless Power Transfer
Links

 Springer

Pablo Pérez-Nicoli
Universidad de la República
Facultad de Ingeniería
Montevideo, Uruguay

Fernando Silveira
Universidad de la República
Facultad de Ingeniería
Montevideo, Uruguay

Maysam Ghovanloo
Bionic Sciences, Inc.
Atlanta, GA, USA

Explore the ideas in the book and test them in the context of your interest with "hands-on" calculation scripts and schematics ready for simulation https://www.springer.com/us/book/9783030654764.

ISBN 978-3-030-65479-5   ISBN 978-3-030-65477-1   (eBook)
https://doi.org/10.1007/978-3-030-65477-1

This Springer imprint is published by the registered company Springer Nature Switzerland AG
The registered company address is: Gewerbestrasse 11, 6330 Cham, Switzerland

Preface

The fundamental physics behind wireless power transfer (WPT) was discovered and evolved in the nineteenth century. However, besides some groundbreaking work by scientists and technologists, like Nikola Tesla, it did not find widespread application in our daily lives until very recently. Today, WPT is an important technique that is rapidly finding applications in new areas from consumer electronics to medical devices and electric vehicles. Although WPT is already commercially used in various applications, challenges remain, and there is room for improvements in many existing and new designs, which are pursued in a large volume of scientific and technical literature that continue being published on the design, analysis, optimization, and applications of WPT links. Most of this literature, which is in the form of academic papers, is focused on a particular part of the wireless link without presenting a general overview of the entire system with its interacting components from the energy source on one end to the power consuming load on the other end. Additionally, in many cases, they tend to come up with different approaches to solve the same problems. Each research group addresses the analysis based on their background and research focus. Therefore, fundamental and fairly simple concepts, such as the WPT link efficiency, can be found in the literature expressed in terms of different parameters, which makes it difficult to grasp not only for those who are entering the field but also for engineers with considerable experience. Physicists have addressed the inductive link as a coupled resonator system and expressed the efficiency in terms of intrinsic decay rates and coupling coefficients, while radio-frequency engineers prefer to use the scattering (S) parameters borrowed from communication systems that often operate at much higher frequencies. Another approach is to deduce the equivalent circuit, using reflected load theory, as we do here in this book, which is both intuitive for engineers and comprehensive to be applicable to various scenarios. It is also the most suitable, in our opinion, to address inductive WPT links, also known as near-field WPT links, which is the focus of this book. Even works that use reflected load theory differ in the way efficiency is presented. Because efficiency can be expressed in terms of self and mutual inductances, or in terms of the coils' quality factors and a normalized parameter, called the coupling coefficient. This makes it difficult for researchers, designers, and graduate students to grasp the fundamental concepts to be able to choose between different approaches in a comprehensive fashion by considering the entire system, which is a necessary step towards optimization of a design for a certain application. They usually lack tangible methods to guide their designs ahead of running iterative computer simulations in a way that would be both practical, efficient, and optimal.

This book was built upon the research work carried on by Maysam Ghovanloo and his team. Their approach, using reflected load theory and expressing the link performance in terms of the coils' quality factors and the coupling coefficient, proved to be both intuitive and a powerful modeling tool. Yet those were dispersed in several academic publications. Those concepts were then adopted, generalized, and applied to more industrial applications, such as radio frequency identification (RFID) in the Ph.D. work of Pablo Pérez-Nicoli under the supervision of Fernando Silveira at Universidad de la República, Uruguay. After the conclusion of his Ph.D. work, which was passed with flying colors with Maysam and a few other WPT experts in the Jury, the need and the opportunity for writing a book from scratch that would allow the reader to gain deep insight into the fundamentals of WPT, the trade-offs involved, and the foundations of the advanced techniques in the field became quite clear among the co-authors. That is how this book was conceived to bring together a vast body of knowledge to address WPT from a rigorous theoretical approach, including numerous examples, further supported by additional material in the form of LTspice and Matlab/Octave scripts, plus the step-by-step deduction of each theoretical expression either in the text or in the appendixes. Our goal was providing fundamental and yet easy to grasp and follow guidelines that would help the design of high-efficiency inductive WPT links. The result of this effort is a book that covers both introductory and advanced levels of theory and practice, providing the reader with enough background knowledge and tools to carry on from simple to advanced WPT concepts in the broader context of system design. We are satisfied with the result and hope that this work will contribute to not only train new designers in this rapidly growing field but also to support experts with ample hands-on experience who may feel that their knowledge of WPT fundamentals could be further strengthened.

The road was not easy, but we had the help and valuable advice of many colleagues who contributed by discussing various topics as well as reviewing the book chapters. We also appreciate the support of the very professional and efficient staff at Springer. More specifically, we would like to thank the reviewers: Nicolas Gammarano and Gonzalo Cuñarro from Universidad de la República, Uruguay; Prof. Ulkuhan Guler from Worcester Polytechnic Institute (WPI), MA, USA; Prof. Yaoyao Jia from North Carolina State University (NCSU), NC, USA; and Dr. Pengcheng Zhang from Hebei University of Technology, Tianjin, China. Additionally, we would like to thank those who contributed with valuable discussions about different topics on WPT: Prof. Alessandra Costanzo from Università di Bologna, Italy, Prof. Alfredo Arnaud from Universidad Católica del Uruguay, Prof. Lionel Pichon and his team from Laboratoire de Génie Electrique de Paris, Prof. Juan Pablo Oliver, Dr. Francisco Veirano and Pedro Arzuaga from Universidad de la República Uruguay, and Federico de Mula and Gabriel Barbat from Integer Montevideo. We wish to acknowledge the financial support to research works at Universidad de la República, Uruguay, by CAP, CSIC, and ANII. Needless to say that our families were also quite supportive over the course of this collaborative work and deserve our sincere appreciation.

Last but not least, we would like to thank you, the reader, who gives meaning to this book, and encourage you to give us feedback and help us in further improving the contents of this book in its future editions for the benefit of many other readers. We hope this book and the additional supporting material help you to understand WPT from the basic concepts to advanced designs and optimization techniques while enjoying it!

Montevideo, Uruguay   Pablo Pérez-Nicoli

Montevideo, Uruguay   Fernando Silveira

Atlanta, USA   Maysam Ghovanloo

Contents

Acronyms

AC	Alternating Current.
AIMD	Active Implantable Medical Device.
BTE	Behind-the-Ear.
CCM	Continuous Conduction Mode.
CENELEC	European Committee for Electrotechnical Standardization.
COTS	Commercially-Available Off-the-Shelf.
DC	Direct Current.
DCM	Discontinuous Conduction Mode.
EMC	Electromagnetic Compatibility.
ESR	Equivalent Series Resistance.
ETSI	European Telecommunications Standards Institute.
EV	Electric Vehicle.
FDA	Food and Drug Administration.
FCC	Federal Communications Commission.
FEA	Finite Element Analysis.
FM	Frequency Modulation.
FSK	Frequency Shift Keying.
ICNIRP	International Commission on Non-Ionizing Radiation Protection.
IEC	International Electrotechnical Commission.
IEEE	Institute of Electrical and Electronic Engineers.

ISI	Intersymbol Interference.
ISM	Industrial, Scientific and Medical.
LDO	Low-Dropout Regulator.
LSK	Load Shift Keying.
MEP	Maximum Efficiency Point.
MPP	Maximum Power Point.
OOP	Optimum Operating Point.
PCE	Power Conversion Efficiency.
PDL	Power Delivered to the Load.
PDM	Pulse-Density Modulation.
PLL	Phase-Locked Loop.
PSC	Printed Spiral Coils.
PSM	Pulse-Shift Modulation.
PTE	Power Transfer Efficiency.
PWM	Pulse-Width Modulation.
P&O	Perturbation and Observation.
RF	Radio Frequency.
RFID	Radio Frequency Identification.
SAR	Specific Absorption Rate.
SRF	Self-Resonant Frequency.
VCR	Voltage Conversion Ratio.
WPT	Wireless Power Transfer.
Rx	Receiver.
Rx-circuit	Receiver Circuit.
Tx	Transmitter.
Tx-circuit	Transmitter Circuit.
f	Carrier frequency of the power transfer.
ω	Carrier angular frequency of the power transfer, $\omega = 2\pi f$.
L_{Tx}	Transmitter coil self-inductance.

R_{Tx}	Transmitter coil parasitic series resistance.
Q_{Tx}	Transmitter coil quality factor $= \frac{\omega L_{Tx}}{R_{Tx}}$.
C_{Tx}	Capacitance added in the transmitter (usually to achieve resonance).
L_{Rx}	Receiver coil self-inductance.
R_{Rx}	Receiver coil parasitic series resistance.
Q_{Rx}	Receiver coil quality factor $= \frac{\omega L_{Rx}}{R_{Rx}}$.
C_{Rx}	Capacitance added in the receiver (usually to achieve resonance).
L_A	Resonator (additional coil) self-inductance.
R_A	Resonator (additional coil) parasitic series resistance.
Q_A	Resonator (additional coil) quality factor $= \frac{\omega L_A}{R_A}$.
$k_{Tx\text{-}Rx}$	Coupling coefficient between the transmitter and the receiver.
$M_{Tx\text{-}Rx}$	Mutual inductance between the transmitter and the receiver.
$D_{Tx\text{-}Rx}$	Distance between the transmitter and the receiver.
$k_{Tx\text{-}A}$	Coupling coefficient between the transmitter and the resonator (additional coil).
$k_{A\text{-}Rx}$	Coupling coefficient between the resonator (additional coil) and the receiver.
R_L	Load resistance, it models the device being powered.
$R_{DC\text{-}DC}$	Receiver DC-DC input resistance.
R_{rect}	Rectifier input resistance, it is equal to the Rx matching network load resistance. See Fig. 6.1.
Z_{MN}	Load impedance of the receiver coil, see Fig. 1.3.
$Z_{MN_{opt\text{-}\eta}}$	Optimum value for Z_{MN} that maximizes the link efficiency η_{Link}.
$Z_{MN_{opt\text{-}P_{MN}}}$	Optimum value for Z_{MN} that maximizes the power delivered to the Rx-circuit, P_{MN}.
$Z_{MN_{opt}}$	Optimum value of Z_{MN}, it refers to both $Z_{MN_{opt\text{-}\eta}}$ and $Z_{MN_{opt\text{-}P_{MN}}}$.
Q_L	Load quality factor, $Q_L = \omega L_{Rx}/\text{Re}\{Z_{MN}\}$.
$Q_{L_{opt\text{-}\eta}}$	Optimum value for Q_L that maximizes the link efficiency, η_{Link}.
$Q_{L_{opt\text{-}P_{MN}}}$	Optimum value for Q_L that maximizes the power delivered to the Rx-circuit, P_{MN}.
$Q_{Rx\text{-}L}$	Rx coil equivalent quality factor, $Q_{Rx\text{-}L} = Q_{Rx}Q_L/(Q_{Rx}+Q_L) = \omega L_{Rx}/(R_{Rx}+\text{Re}\{Z_{MN}\})$.
V_L	Load voltage, see Fig. 1.3.

V_S — Voltage of the power source in the transmitter, see Fig. 1.3.

V_{rect} — Rectifier output voltage, see Fig. 1.3. It is equal to V_L if there is no DC-DC converter.

i_{Tx} — Peak current through Tx coil.

i_{Rx} — Peak current through Rx coil.

P_L — Power delivered to the load (DC), see Fig. 1.3.

$P_{DC\text{-}DC}$ — Receiver DC-DC converter input power.

P_{rect} — Receiver rectifier input power.

P_{MN} — Rx-circuit input power, see Fig. 1.3.

P_{Tx} — Tx-circuit output power, see Fig. 1.3.

P_S — Output power of the source in the transmitter, see Fig. 1.3.

η_{Link} — Inductive link efficiency, see Fig. 1.3.

η_{MN} — Receiver matching network efficiency, see Fig. 7.6.

η_{rect} — Receiver rectifier efficiency, see Fig. 7.6.

$\eta_{DC\text{-}DC}$ — Receiver DC-DC converter efficiency, see Fig. 7.6.

η_{Rx} — Receiver circuit efficiency, see Fig. 1.3 ($\eta_{Rx} = \eta_{MN} \cdot \eta_{rect} \cdot \eta_{DC\text{-}DC}$).

η_{Tx} — Tx-circuit efficiency, see Fig. 1.3.

η_{TOT} — Total system efficiency $\eta_{TOT} = P_L/P_S$, see Fig. 1.3 ($\eta_{TOT} = \eta_{Tx} \cdot \eta_{Link} \cdot \eta_{Rx}$).

G_{rect} — Rectifier gain defined as the ratio between the rectifier output voltage, V_{rect}, and its input peak voltage, $V_{in_{peak}}$, $G_{rect} = \frac{V_{rect}}{V_{in_{peak}}}$.

$G_{DC\text{-}DC}$ — Receiver DC-DC converter gain defined as the ratio between its output voltage, V_L, and its input voltage, V_{rect}, $G_{DC\text{-}DC} = \frac{V_L}{V_{rect}}$.

$G_{Tx_{DC\text{-}DC}}$ — Transmitter DC-DC converter gain defined as the ratio between its output voltage and its input voltage, V_S.

$\mathbb{K}_{Tx\text{-}A}$ — $\mathbb{K}_{Tx\text{-}A} = k_{Tx\text{-}A}{}^2 Q_{Tx} Q_A$.

$\mathbb{K}_{A\text{-}Rx}$ — $\mathbb{K}_{A\text{-}Rx} = k_{A\text{-}Rx}{}^2 Q_A Q_{Rx}$.

$\mathbb{K}_{A\text{-}Rx\text{-}L}$ — $\mathbb{K}_{A\text{-}Rx\text{-}L} = k_{A\text{-}Rx}{}^2 Q_A Q_{Rx\text{-}L}$.

$\mathbb{K}_{Tx\text{-}Rx}$ — $\mathbb{K}_{Tx\text{-}Rx} = k_{Tx\text{-}Rx}^2 Q_{Tx} Q_{Rx}$.

$\mathbb{K}_{Tx\text{-}Rx\text{-}L}$ — $\mathbb{K}_{Tx\text{-}Rx\text{-}L} = k_{Tx\text{-}Rx}^2 Q_{Tx} Q_{Rx\text{-}L}$.

Chapter 1
Introduction to Wireless Power Transfer

1.1 Why Wireless?

In the past decade, we have witnessed a dramatic increase in the number of mobile devices which are used in a wide range of applications and contexts. Either primary or rechargeable batteries are often used as energy source or storage elements to power them. For applications such as wireless sensors, or Active Implantable Medical Devices (AIMDs) [1–3], replacing the batteries may be impractical, expensive, risky, or in some cases impossible. To recharge these batteries, the power can be harvested by the mobile devices themselves, but in many applications this power may not be enough. Wireless Power Transfer (WPT) to these devices is a favorable solution to either recharge the batteries, as shown in Fig. 1.1a, or avoid them altogether. Furthermore, WPT is also used in several other applications, as mentioned in the following.

The most popular application, in which WPT substitutes batteries, is the Radio Frequency Identification (RFID), an example of which is shown in Fig. 1.1b [4, 5]. Passive RFID tags that do not rely on batteries are more robust and cheaper and have a longer lifetime. Nowadays, WPT is also used to power or recharge devices that have traditionally used a power cord, such as mobile phones (Fig. 1.1c) [6, 7], electric cars (Fig. 1.1d) [8, 9], and home appliances (Fig. 1.1e) [10, 11]. In these applications, WPT not only is a practical solution but also enables new possibilities like electric recharging lanes among many others. A number of studies have focused on long-distance and large-power WPT systems [12]. These systems will enable futuristic applications like those proposed in [13], where solar power is collected in the space and transferred wirelessly to the earth (Fig. 1.1f).

The main parameters that define the WPT link are (1) the distance, $D_{Tx\text{-}Rx}$, between Transmitter (Tx) and Receiver (Rx), and their relative orientation with respect to one another; (2) the Tx and Rx coils areas, A_{Tx} and A_{Rx} respectively; (3) the power carrier frequency, f, (4) the Power Delivered to the Load (PDL), P_L;

(a)   **(b)**   **(c)**

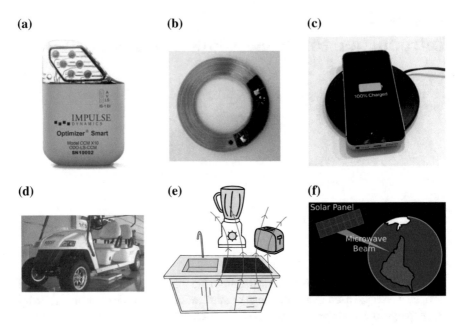

(d)   **(e)**   **(f)**

Fig. 1.1 Wireless power transfer applications. (**a**) AIMDs. (Image Credit: Impulse Dynamics. Used with permission). (**b**) RFID. (©2016 IEEE Reprinted, with permission, from [5]). (**c**) Phones. (**d**) Vehicles. (©2016 IEEE Reprinted, with permission, from [8]). (**e**) Home appliances. (**f**) Solar power satellite

and (5) Power Transfer Efficiency (PTE), η_{TOT}, defined as the PDL over the power taken from the primary power supply.

Many trade-offs exist between these parameters, and the optimization of these parameters depends on the application. For Electric Vehicle (EV) applications, constraints on the Tx and Rx coils' sizes are relatively relaxed, and the main focus is on achieving high PTE at sufficient PDL to minimize heat dissipation while reducing the weight of the Rx coil. On the other hand, in AIMDs, the Rx size is one of the main constraints, and the power levels are orders of magnitude lower than those for EV, while the dissipated heat in both Tx and Rx coils and power management circuits should be strictly limited to prevent excessive temperature elevation, which may harm the surrounding human tissue.

1.2 Wireless Links Classifications

There are various classifications of WPT systems, but typically they are divided into near-field and far-field links. This classification is based on their physical working principle; in the far-field, an electromagnetic wave transports the energy, while near-field relies on magnetic or electric coupling. For antennas much shorter than the

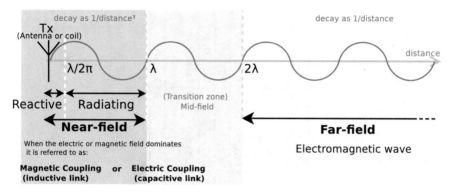

Fig. 1.2 Near- and far-field approximated regions for antennas much shorter than the wavelength, λ, [14, 15]. The WPT links are usually classified based on these operating regions. Examples of different links in the literature are presented in Table 1.1

wavelength, λ, the near-field and far-field zones can be approximated as in Fig. 1.2. More precise boundaries can be defined based primarily on the antenna type and antenna size, and even then the experts differ [14]. The region between the near-field and the far-field is called the transition zone or mid-field, which has a combination of the characteristics found in both the near-field and the far-field [14]. The near-field can be sub-divided into magnetic or electric coupling, depending on the dominant field.

To achieve long distances, in the range of meters, far-field is preferred because the beam can be pointed toward the Rx. This beam-based WPT system can transfer large power (kilowatts) at large distances (tens of meters) with high efficiency (>50%) at the risk of interference with other radio signals [12]. However, for short distances (tens of centimeters), higher efficiencies could be achieved using near-field links [16] that do not require a line-of-sight operation.

Although in the near-field region many of the proposed electric coupling links are for short-gap distances (millimeters) due to constraints on the developed voltage [17], larger gaps (≃10 cm) can be achieved [18, 19]. One advantage of using these links is that the power can be transferred through metal barriers. However, links based on the magnetic field (i.e., inductive WPT links) are more common, and many examples of transferring power in the centimeters range have been presented. The magnetic field causes less adverse effects on the human body than the electric field; thus inductive WPT is the best choice for biomedical systems [16].

A detailed description and comparison between different WPT mechanisms, including the optical and ultrasound links, can be found in classical references, such as [20, 21]. In Table 1.1, examples of different links in the literature are presented by the way of summary.

In this book, we focus on the design of inductive (near-field) WPT systems which are further introduced in Sect. 1.3.

Table 1.1 Examples of different types of electromagnetic wireless power transfer links

	Near-field					Far-field	
	Magnetic coupling			Electric coupling			
	[22]	[23]	[8]	[19]	[24]	[12]	[25]
Main application	n/a	AIMD	EV	EV	Smartphones	Aerospace	IoT
Distance	20 cm	3.3 cm	15.6 cm	12 cm	0.013 cm	1.6 km	1.2 m
Rx area	78.5 cm²	6.2 cm²	189 cm²	150 cm²	6 cm²	24.5 m²	4.5 mm²
f	4 MHz	1 MHz	20.15 kHz	13.56 MHz	4.2 MHz	2.388 GHz	907 MHz
Vacuum wavelength $\lambda = \frac{c}{f}$	75 m	300 m	12 km	22 m	71 m	126 mm	331 mm
P_L	2 W	20 mW	1000 W	884 W	3.7 W	30 kW	10 mW
η (%)	45	65	96	91	80	8	0.5

1.3 Inductive Wireless Power Transfer

A general block diagram of an inductive WPT system is presented in Fig. 1.3. The Transmitter Circuit (Tx-circuit) generates an alternating current in the Tx coil, labeled L_{Tx}, which induces an alternating voltage in the Rx coil, labeled L_{Rx}. The Receiver Circuit (Rx-circuit) adapts this induced voltage to power the load, R_L. The total system efficiency, η_{TOT}, is determined by the link efficiency, η_{Link}, the Tx-circuit efficiency, η_{Tx}, and the Rx-circuit efficiency, η_{Rx}. Figure 1.3 also considers the possibility to use additional resonant coils placed between the Tx and Rx coils, which is further addressed in Sect. 1.3.4.

1.3.1 Transmitter DC-DC Converter

Every WPT system includes an energy source on the Tx side, which, depending on whether the system is meant to be portable or not, is a battery pack or wall-plugged power supply. The output voltage of this source, however, may not be optimal for the system at all times. Therefore a DC-DC power converter is often used following the energy source to control the transmitted power, P_{Tx} in Fig. 1.3 [26–29]. By controlling the transmitted power in a closed-loop, the efficiency of the link can be improved because instead of transmitting power based on the worst-

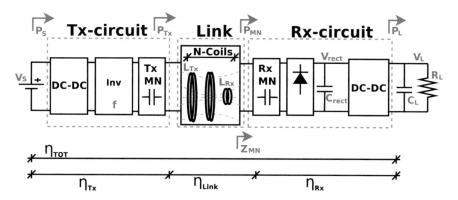

Fig. 1.3 Block diagram of an inductive WPT system. Acronyms: DC-DC, DC-to-DC converter; Inv, inverter (DC-to-AC converter). L_{Tx} and L_{Rx} are the Tx and Rx coils' self-inductances, respectively. P_S, P_{Tx}, P_{MN}, and P_L are the power delivered at the terminals indicated in the figure. $\eta_{Tx} = P_{Tx}/P_S$, $\eta_{Link} = P_{MN}/P_{Tx}$, and $\eta_{Rx} = P_L/P_{MN}$ are the Tx-circuit, link, and Rx-circuit efficiencies, respectively. $\eta_{TOT} = P_L/P_S$ is the total efficiency. V_S is the voltage of the power source in the Tx and V_L is the load voltage in the Rx. R_L models the load circuit. C_{rect} and C_L are low-pass filter and load capacitors, respectively. The Tx-MN and Rx-MN represent the Tx and Rx matching networks, respectively. Most of these terms are widely used throughout the book; thus they are defined in the glossary

case scenario to ensure that the Rx remains functional, the transmitted power can be automatically adjusted based on how much power is needed and actually received on the Rx side. The analysis of the closed-loop that controls the transmitted power is addressed in Chap. 7. Additionally, an open-loop link with too much output power may damage the Rx when it is too close to the Tx and may also generate undesired heat dissipation in the Tx, the Rx, or the surrounding foreign objects (e.g., human tissue in AIMDs). It should be mentioned that, as is discussed in Sect. 1.3.2, the transmitted power can also be controlled by the inverter.

The DC-DC converter can be a step-up, a step-down, or both step-up and step-down, depending on the battery voltage, the inverter architecture used, and the desired transmitted power. Although the heat dissipation in the Tx is generally not as much a concern as the heat dissipation in the Rx (e.g., in AIMDs), high efficiency is desired as in any other block of Fig. 1.3 to extend the battery life.

This converter must be able to variate its output voltage within the range required by the inverter in order to deliver the desired transmitted power. It should be considered that the inverter will consume a different current (power) for each DC-DC converter output voltage, which should be taken into account during the converter design.

1.3.2   Inverter

Since DC power does not pass through an inductive WPT link (due to Faraday law, see Chap. 2), an inverter is needed in the power flow. This inverter is also where the AC power carrier is generated and its frequency is determined.

Theoretically, a class-A or class-B power amplifier could be used to drive the Tx coil [30], but they achieve low efficiencies [20]. The class-C power amplifier has been used in some previous works [31, 32]. However, the most efficient and thus typically used architectures are class-D [27, 28, 33] and class-E [26, 29, 34–37] inverters. A comparison between these different architectures can be found in classical references, such as [20, 38].

The inverter efficiency and output power depend on its load impedance. Therefore, the Tx matching network, addressed in Sect. 1.3.3, is used to adjust the inverter load impedance. As it is further analyzed in Chap. 7, changes in the coupling between the Tx and the Rx coils alter the inverter load impedance, affecting the efficiency, especially for class-E amplifiers. Therefore, the load impedance range should be considered during the inverter design.

The transmitted power, P_{Tx}, can be controlled directly in the inverter, controlling the on/off of the switching element, e.g., adjusting its frequency or duty cycle. In [33], the output power of a class-D inverter is controlled through switching-frequency modulation, while in [34] a class-E inverter is turned on and off to control the mean output power.

1.3.3 Tx Matching Network

As was mentioned in Sect. 1.3.2, the Tx matching network is used to adapt the load impedance of the inverter [36]. Usually, the Tx matching network is designed to achieve resonance in the Tx by canceling the Tx coil reactance. However, a non-resonant Tx could be useful to limit the inverter output current. Additionally, in this book we present how the resonance or lack thereof in the Tx coil affects the closed-loop that controls the transmitted power in Chap. 7.

Any change in the distance or alignment between coils affects their coupling coefficient, thus altering the inverter load impedance. To dynamically adjust this load impedance, an adaptive matching network was proposed in [39–41]. However, fixed matching networks are typically used, and the Tx-circuit is designed to bear the expected coupling variations.

1.3.4 Inductive Link

The inductive link works basically as a transformer, transferring the power through the magnetic field, which is alternating at the carrier frequency. This block is further analyzed throughout the book, but especially in Chaps. 2 and 3. Therefore, in this section, we only present an overview of the efficiency achieved by these links.

In systems where a large distance between relatively small coils is desired, the link efficiency, η_{Link}, is often the one that limits the total PTE, η_{TOT}. The Rx coil is usually the one with more size constraints, for example, in AIMDs. In Fig. 1.4 a few examples from the literature are presented to highlight quantitatively the difficulty in achieving long distance with small receivers. The position of each dot indicates the Rx area and target distance, while the dot color represents the efficiency achieved by the work in grayscale. In the area above the gray curve (long distance with small receiver zone), the efficiency is theoretically less than 1% for a system with the following characteristics: (1) Tx 5 times larger (in area) than the Rx coil, (2) coils with circular planar shape, (3) perfect alignment, (4) $Q_{Tx} = 300$ and $Q_{Rx} = 60$, and (5) optimum load condition (this last point is addressed in Chap. 5).

To overcome this efficiency limitation, Kurs et al. [52] proposed a novel magnetic link using additional resonant coils which increases the efficiency for a given distance, or the power transfer distance for a given efficiency. Although authors in [52] proposed a 4-coil link and it was used by many others, such as [29, 53], the same principle can be used to build a 3-coil link [50, 54]. On the other hand, other works extend this idea to generate an N-relay coils link [55]. As shown in Fig. 1.5, in comparison with Fig. 1.4, systems with one additional coil (3-coil links) can achieve higher efficiencies at larger distances even with small Rx coils, in comparison with the 2-coil link. The location of the additional resonant coils depends on the application, in [49] a 3-coil WPT link for millimeter-sized biomedical implants is implemented implanting the additional resonator close to the Rx, while in [51],

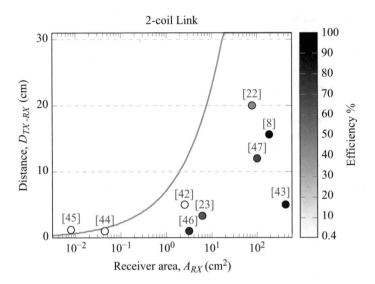

Fig. 1.4 State-of-the-art in 2-coil links: distance, Rx coil area, and link efficiency, η_{Link}. In this figure, we include relevant papers that achieve the highest efficiencies and largest distances. The solid curve delimits the zone where the efficiency is theoretically less than 1% for a system with the following characteristics: (1) Tx five times larger (in area) than the Rx coil, (2) coils with circular planar shape, (3) perfect alignment, (4) $Q_{Tx} = 300$ and $Q_{Rx} = 60$, and (5) optimum load condition

for a motion-free endoscopy capsule, the 3-coil link is implemented placing the additional resonator on the user jacket.

In Fig. 1.5 the gray line is delimiting the zone with less than 1% efficiency for a system with the following characteristics: (1) additional resonator equal to the Tx and both five times larger (in area) than the Rx coil; (2) the additional resonator is placed $0.9 \times D_{Tx-Rx}$ from the Tx (thus near the Rx); (3) the three coils have a circular planar shape; (4) perfect alignment; (5) $Q_{Tx} = Q_A = 300$ where Q_A is the quality factor of the additional resonator; (6) $Q_{Rx} = 60$; and (7) optimum load condition (this last point is addressed in Chap. 5).

In this book, we address the inductive link efficiency especially considering the use of additional resonant coils in Chap. 2, the practical aspects in the coil design in Chap. 3, and how to achieve the optimum operating point (optimum load condition) in Chaps. 5, 6, and 7. Additionally, the link design is taken into account from a system-level perspective, considering, for instance, the effect that the use of the link to transmit data from the Rx to the Tx (back telemetry) may have on it.

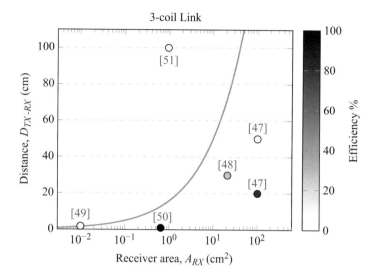

Fig. 1.5 State-of-the-art in 3-coil links: distance, Rx coil area, and link efficiency, η_{Link}. In this figure, we include relevant papers that achieve the highest efficiencies and largest distances. The solid curve delimits the zone with less than 1% efficiency for a system with the following characteristics: (1) additional resonator equal to the Tx and both five times larger (in area) than the Rx coil, (2) the additional resonator is placed $0.9 \times D_{Tx\text{-}Rx}$ from the Tx (thus near the Rx), (3) the three coils have a circular planar shape, (4) perfect alignment, (5) $Q_{Tx} = Q_A = 300$ where Q_A is the quality factor of the additional resonator, and (6) $Q_{Rx} = 60$, (7) optimum load condition

1.3.5 Rx Matching Network

The Rx matching network is a key block in any WPT system. Its main goal is to adapt the load impedance of the Rx coil with the rest of the inductive link. Usually, it is used to cancel the Rx coil reactance, i.e., to achieve resonance. However, the real part can also be adapted to maximize the link efficiency. The optimum Rx coil load condition and this block (Rx matching network) are further analyzed in Chap. 5.

This block has received different names such as resonant/resonance capacitor/capacitance [20, 35, 43, 48, 50, 56], resonant structure [22], resonant load transformation [42], compensation capacitances [57], matching capacitors [8, 27, 36], and matching network [39, 41, 58]. These denominations depend on the authors and how the block is implemented, e.g., with one capacitor, two capacitors, or an L-C network. Although the name "matching network" is more typically used for higher frequency circuits, we adopt this terminology in this book to highlight that it can be implemented in different ways and not only with one capacitor.

1.3.6   Rectifier

A rectifier is often required following the inductive link because the load circuit usually needs a DC voltage. Its design strongly depends on the operating frequency and power level. Different architectures can be used, full-bridge diode rectifiers [27, 59], passive voltage-multiplier rectifiers [29, 60–63], active diodes or active rectifiers [23, 32, 64–68], class E rectifiers [36, 37], and even LC-based oscillator structures [69].

Regardless of the selected architecture, the goal is to rectify the input signal as efficiently as possible, minimizing power dissipation and sometimes regulating the output voltage as well [33].

Reconfigurable rectifier architectures were proposed to achieve higher power transfer and provide dynamic impedance matching [23, 33, 68]. The analysis of these reconfigurable architectures and how they affect the WPT link is addressed in Chap. 6.

Communication from the Rx-circuit to the Tx-circuit is needed in many cases in order to adjust the Tx power. This is sometimes called back telemetry and could also be used as a general-purpose data communication channel in the system. One approach is the so-called backscattering, which is based on detecting the changes in the Rx load through the Tx (load shifting). Load shifting is usually implemented with a shunt switch, which short-circuits the rectifier input nodes. The reconfigurable architecture also results in a load shift that can be used for backscattering [33]. The use of active rectifier in the back telemetry is addressed in Chap. 4.

1.3.7   Receiver DC-DC Converter

A DC-DC converter can be placed after the rectifier to adapt/regulate its output voltage to the value required by the load. This is not the only approach to regulate the output voltage, because the transmitted power can also be regulated (as mentioned in Sect. 1.3.1) for this purpose. Different approaches for output voltage regulation are addressed in Chap. 7.

The most common architectures used for this block are Low-Dropout Regulators (LDOs) [23, 26], boost converters [60], buck converters [34], buck-boost converters [70, 71], and switched-capacitor converters [72, 73]. As all the other blocks, high efficiency is required to take advantage of all the received power and avoid heat dissipation. Other key factors are line regulation factor, load regulation factor, and the ability to reduce high-frequency ripple on the power supply.

When the DC-DC converter changes its gain for regulation, it affects the rectifier load impedance, which ends up affecting the Rx coil load impedance, Z_{MN} in Fig. 1.3, and the Rx coil matching. The same occurs when a reconfigurable rectifier architecture is used, as mentioned in Sect. 1.3.6. This effect and the design to continuously track and operate in the optimum operating point are analyzed in depth in Chap. 6.

References

1. R.R. Harrison, Designing efficient inductive power links for implantable devices, in *IEEE International Symposium on Circuits and Systems* (2007), pp. 2080–2083. https://doi.org/10.1109/ISCAS.2007.378508
2. T. Campi, S. Cruciani, F. Palandrani, V. De Santis, A. Hirata, M. Feliziani, Wireless power transfer charging system for AIMDs and pacemakers. IEEE Trans. Microw. Theory Techn. **64**(2), 633–642 (2016)
3. B.H. Waters, A.P. Sample, P. Bonde, J.R. Smith, Powering a ventricular assist device (VAD) with the free-range resonant electrical energy delivery (FREE-D) system. Proc. IEEE **100**(1), 138–149 (2012). ISSN 1558-2256. https://doi.org/10.1109/JPROC.2011.2165309
4. A.J. Soares Boaventura, N. Carvalho, Extending reading range of commercial RFID readers. IEEE Trans. Microw. Theory Techn. **61**(1), 633–640 (2013). ISSN 0018-9480. https://doi.org/10.1109/TMTT.2012.2229288
5. P. Pérez-Nicoli, A. Rodríguez-Esteva, F. Silveira, Bidirectional analysis and design of RFID using an additional resonant coil to enhance read range. IEEE Trans. Microw. Theory Techn. **64**(7), 2357–2367 (2016). ISSN 0018-9480. https://doi.org/10.1109/TMTT.2016.2573275
6. C.-G. Kim, D.-H.Seo, J.-S. You, J.-H. Park, B.H. Cho, Design of a contactless battery charger for cellular phone. IEEE Trans. Ind. Electron. **48**(6), 1238–1247 (2001). ISSN 1557-9948. https://doi.org/10.1109/41.969404
7. J. Zhu, Y. Ban, Y. Zhang, Z. Yan, R. Xu, C.C. Mi, Three-coil wireless charging system for metal-cover smartphone applications. IEEE Trans. Power Electron. **35**(5), 4847–4858 (2020). ISSN 1941-0107. https://doi.org/10.1109/TPEL.2019.2944845
8. H. Kim, C. Song, D.H. Kim, D.H. Jung, I.M. Kim, Y.I. Kim, J. Kim, S. Ahn, J. Kim, Coil design and measurements of automotive magnetic resonant wireless charging system for high-efficiency and low magnetic field leakage. IEEE Trans. Microw. Theory Techn. **64**(2), 383–400 (2016). ISSN 0018-9480. https://doi.org/10.1109/TMTT.2015.2513394
9. J. Shin, S. Shin, Y. Kim, S. Ahn, S. Lee, G. Jung, S.-J. Jeon, D.-H. Cho, Design and implementation of shaped magnetic-resonance-based wireless power transfer system for roadway-powered moving electric vehicles. IEEE Trans. Ind. Electron. **61**(3), 1179–1192 (2013)
10. M. Itraj, W. Ettes, Topology study for an inductive power transmitter for cordless kitchen appliances, in *IEEE PELS Workshop on Emerging Technologies, Wireless Power Transfer (WoW)* (2018), pp. 1–8. https://doi.org/10.1109/WoW.2018.8450898
11. A.M. Jawad, R. Nordin, S.K. Gharghan, H.M. Jawad, M. Ismail, M.J. Abu-AlShaeer, Single-tube and multi-turn coil near-field wireless power transfer for low-power home appliances. Energies **11**(8), 1969 (2018)
12. N. Shinohara, History of research and development of beam wireless power transfer, in *IEEE Wireless Power Transfer Conference* (IEEE, 2018), pp. 1–3
13. P.E. Glaser, Power from the sun: its future. Science **162**(3856), 857–861 (1968)
14. OSHA-Cincinnati-Laboratory, Electromagnetic radiation: field memo, electromagnetic radiation and how it affects your instruments. Technical report, United States Department of Labor, 1990
15. P.V. Nikitin, K.V.S. Rao, S. Lazar, An overview of near field UHF RFID, in *IEEE International Conference on RFID* (2007), pp. 167–174. https://doi.org/10.1109/RFID.2007.346165
16. T. Sun, X. Xie, Z. Wang, *Wireless Power Transfer for Medical Microsystems* (Springer-Verlag, New York, 2013)
17. J. Dai, D.C. Ludois, A survey of wireless power transfer and a critical comparison of inductive and capacitive coupling for small gap applications. IEEE Trans. Power Electron. **30**(11), 6017–6029 (2015). ISSN 0885-8993. https://doi.org/10.1109/TPEL.2015.2415253
18. S. Sinha, B. Regensburger, K. Doubleday, A. Kumar, S. Pervaiz, K.K. Afridi, High-power-transfer-density capacitive wireless power transfer system for electric vehicle charging, in *IEEE Energy Conversion Congress and Exposition* (2017), pp. 967–974. https://doi.org/10.1109/ECCE.2017.8095890

19. B. Regensburger, A. Kumar, S. Sinha, K. Afridi, High-performance 13.56-MHz large air-gap capacitive wireless power transfer system for electric vehicle charging, in *IEEE Workshop on Control and Modeling for Power Electronics* (2018), pp. 1–4. https://doi.org/10.1109/COMPEL.2018.8460153

20. K. Van Schuylenbergh, R. Puers, *Inductive Powering: Basic Theory and Application to Biomedical Systems* (Springer, 2009)

21. Y. Lu, W.-H. Ki, *CMOS Integrated Circuit Design for Wireless Power Transfer* (Springer, 2018)

22. L. Chen, S. Liu, Y.C. Zhou, T.J. Cui, An optimizable circuit structure for high-efficiency wireless power transfer. IEEE Trans. Ind. Electron. **60**(1), 339–349 (2013). ISSN 0278-0046. https://doi.org/10.1109/TIE.2011.2179275

23. H. Lee, An auto-reconfigurable $2 \times /4\times$ AC-DC regulator for wirelessly powered biomedical implants with 28% link efficiency enhancement. IEEE Trans. VLSI Syst. **24**(4), 1598–1602 (2016). ISSN 1063-8210. https://doi.org/10.1109/TVLSI.2015.2452918

24. M. Kline, I. Izyumin, B. Boser, S. Sanders, Capacitive power transfer for contactless charging, in *IEEE Annual Applied Power Electronics Conference and Exposition* (2011), pp. 1398–1404. https://doi.org/10.1109/APEC.2011.5744775

25. W. Lin, R.W. Ziolkowski, J. Huang, Electrically small, low-profile, highly efficient, Huygens dipole rectennas for wirelessly powering internet-of-things devices. IEEE Trans. Antennas Propag. **67**(6), 3670–3679 (2019). ISSN 0018-926X. https://doi.org/10.1109/TAP.2019.2902713

26. G. Wang, W. Liu, M. Sivaprakasam, G.A. Kendir, Design and analysis of an adaptive transcutaneous power telemetry for biomedical implants. IEEE Trans. Circuits Syst. I **52**(10), 2109–2117 (2005). ISSN 1549-8328. https://doi.org/10.1109/TCSI.2005.852923

27. T.D. Yeo, D. Kwon, S.T. Khang, J.W. Yu, Design of maximum efficiency tracking control scheme for closed-loop wireless power charging system employing series resonant tank. IEEE Trans. Power Electron. **32**(1), 471–478 (2017)

28. H. Li, Y. Tang, K. Wang, X. Yang, Analysis and control of post regulation of wireless power transfer systems, in *IEEE Annual Southern Power Electronics Conference* (2016), pp. 1–5. https://doi.org/10.1109/SPEC.2016.7846093

29. C. Yang, C. Chang, S. Lee, S. Chang, L. Chiou, Efficient four-coil wireless power transfer for deep brain stimulation. IEEE Trans. Microw. Theory Techn. **65**(7), 2496–2507 (2017). ISSN 0018-9480. https://doi.org/10.1109/TMTT.2017.2658560

30. B. Hansen, K. Aabo, J. Bojsen, An implantable, externally powered radiotelemetric system for long-term ECG and heart-rate monitoring. Biotelem. Patient Monit. **9**(4), 227–237 (1982)

31. H.M. Lee, M. Ghovanloo, An adaptive reconfigurable active voltage doubler/rectifier for extended-range inductive power transmission. IEEE Trans. Circuits Syst. II **59**(8), 481–485 (2012). ISSN 1549-7747. https://doi.org/10.1109/TCSII.2012.2204840

32. H.M. Lee, M. Ghovanloo, A high frequency active voltage doubler in standard CMOS using offset-controlled comparators for inductive power transmission. IEEE Trans. Biomed. Circuits Syst. **7**(3), 213–224 (2013)

33. X. Li, C.Y. Tsui, W.H. Ki, A 13.56 MHz wireless power transfer system with reconfigurable resonant regulating rectifier and wireless power control for implantable medical devices. IEEE J. Solid-State Circuits **50**(4), 978–989 (2015)

34. M. Fu, H. Yin, M. Liu, C. Ma, Loading and power control for a high-efficiency class E PA-driven megahertz WPT system. IEEE Trans. Ind. Electron. **63**(11), 6867–6876 (2016). ISSN 0278-0046. https://doi.org/10.1109/TIE.2016.2582733

35. B. Lee, P. Yeon, M. Ghovanloo, A multicycle Q-modulation for dynamic optimization of inductive links. IEEE Trans. Ind. Electron. **63**(8), 5091–5100 (2016)

36. M. Liu, S. Liu, C. Ma, A high-efficiency/output power and low-noise megahertz wireless power transfer system over a wide range of mutual inductance. IEEE Trans. Microw. Theory Technol. **65**(11), 4317–4325 (2017). ISSN 0018-9480. https://doi.org/10.1109/TMTT.2017.2691767

37. T. Nagashima, X. Wei, E. Bou, E. Alarcón, M.K. Kazimierczuk, H. Sekiya, Steady-state analysis of isolated class-E^2 converter outside nominal operation. IEEE Trans. Ind. Electron. **64**(4), 3227–3238 (2017). ISSN 0278-0046. https://doi.org/10.1109/TIE.2016.2631439

38. M.K. Kazimierczuk, *RF Power Amplifiers*, vol. 1 (Wiley Online Library, 2008)
39. Y. Lim, H. Tang, S. Lim, J. Park, An adaptive impedance-matching network based on a novel capacitor matrix for wireless power transfer. IEEE Trans. Power Electron. **29**(8), 4403–4413 (2014). ISSN 0885-8993. https://doi.org/10.1109/TPEL.2013.2292596
40. T.C. Beh, M. Kato, T. Imura, S. Oh, Y. Hori, Automated impedance matching system for robust wireless power transfer via magnetic resonance coupling. IEEE Trans. Ind. Electron. **60**(9), 3689–3698 (2013)
41. J. Bito, S. Jeong, M.M. Tentzeris, A novel heuristic passive and active matching circuit design method for wireless power transfer to moving objects. IEEE Trans. Microw. Theory Technol. **65**(4), 1094–1102 (2017). ISSN 0018-9480. https://doi.org/10.1109/TMTT.2017.2672544
42. R.F. Xue, K.W. Cheng, M. Je, High-efficiency wireless power transfer for biomedical implants by optimal resonant load transformation. IEEE Trans. Circuits Syst. I **60**(4), 867–874 (2013). ISSN 1549-8328. https://doi.org/10.1109/TCSI.2012.2209297
43. C. Florian, F. Mastri, R.P. Paganelli, D. Masotti, A. Costanzo, Theoretical and numerical design of a wireless power transmission link with GaN-based transmitter and adaptive receiver. IEEE Trans. Microw. Theory Technol. **62**(4), 931–946 (2014). ISSN 0018-9480. https://doi.org/10.1109/TMTT.2014.2303949
44. M. Zargham, P.G. Gulak, Fully integrated on-chip coil in 0.13 μm CMOS for wireless power transfer through biological media. IEEE Trans. Biomed. Circuits Syst. **9**(2), 259–271 (2015). ISSN 1932-4545. https://doi.org/10.1109/TBCAS.2014.2328318
45. A. Ibrahim, M. Kiani, A figure-of-merit for design and optimization of inductive power transmission links for millimeter-sized biomedical implants. IEEE Trans. Biomed. Circuits Syst. **10**(6), 1100–1111 (2016)
46. M. Schormans, V. Valente, A. Demosthenous, Practical inductive link design for biomedical wireless power transfer: a tutorial. IEEE Trans. Biomed. Circuits Syst. **12**(5), 1112–1130 (2018). ISSN 1932-4545. https://doi.org/10.1109/TBCAS.2018.2846020
47. D. Seo, Comparative analysis of two- and three-coil WPT systems based on transmission efficiency. IEEE Access **7**, 151962–151970 (2019)
48. M. Kiani, M. Ghovanloo, The circuit theory behind coupled-mode magnetic resonance-based wireless power transmission. IEEE Trans. Circuits Syst. I **59**(9), 2065–2074 (2012). ISSN 1549-8328
49. P. Yeon, S.A. Mirbozorgi, M. Ghovanloo, Optimal design of a 3-coil inductive link for millimeter-sized biomedical implants, in *IEEE Biomedical Circuits and Systems Conference (BioCAS)* (2016), pp. 396–399. https://doi.org/10.1109/BioCAS.2016.7833815
50. M. Kiani, U.M. Jow, M. Ghovanloo, Design and optimization of a 3-coil inductive link for efficient wireless power transmission. IEEE Trans. Biomed. Circuits Syst. **5**(6), 579–591 (2011). ISSN 1932-4545. https://doi.org/10.1109/TBCAS.2011.2158431
51. T. Sun, X. Xie, G. Li, Y. Gu, Y. Deng, Z. Wang, A two-hop wireless power transfer system with an efficiency-enhanced power receiver for motion-free capsule endoscopy inspection. IEEE Trans. Biomed. Eng. **59**(11), 3247–3254 (2012). ISSN 0018-9294. https://doi.org/10.1109/TBME.2012.2206809
52. A. Kurs, A. Karalis, R. Moffatt, J.D. Joannopoulos, P. Fisher, M. Soljacic, Wireless power transfer via strongly coupled magnetic resonances. Science **317**(5834), 83–6 (2007). ISSN 1095-9203. https://doi.org/10.1126/science.1143254
53. K. Na, H. Jang, H. Ma, F. Bien, Tracking optimal efficiency of magnetic resonance wireless power transfer system for biomedical capsule endoscopy. IEEE Trans. Microw. Theory Technol. **63**(1), 295–304 (2015). ISSN 0018-9480. https://doi.org/10.1109/TMTT.2014.2365475
54. J. Kim, D.-H. Kim, J. Choi, K.-H. Kim, Y.-J. Park, Free-positioning wireless charging system for small electronic devices using a bowl-shaped transmitting coil. IEEE Trans. Microw. Theory Technol. **63**(3), 791–800 (2015). ISSN 0018-9480. https://doi.org/10.1109/TMTT.2015.2398865
55. F. Zhang, S.A. Hackworth, W. Fu, C. Li, Z. Mao, M. Sun, Relay effect of wireless power transfer using strongly coupled magnetic resonances. IEEE Trans. Magn. **47**(5), 1478–1481 (2011). ISSN 0018-9464. https://doi.org/10.1109/TMAG.2010.2087010

56. S.H. Lee, R.D. Lorenz, Development and validation of model for 95%-efficiency 220-W wireless power transfer over a 30-cm air gap. IEEE Trans. Ind. Appl. **47**(6), 2495–2504 (2011)
57. Y.H. Sohn, B.H. Choi, E.S. Lee, G.C. Lim, G. Cho, C.T. Rim, General unified analyses of two-capacitor inductive power transfer systems: equivalence of current-source SS and SP compensations. IEEE Trans. Power Electron. **30**(11), 6030–6045 (2015). ISSN 0885-8993. https://doi.org/10.1109/TPEL.2015.2409734
58. P. Pérez-Nicoli, F. Silveira, Matching networks for maximum efficiency in two and three coil wireless power transfer systems, in *IEEE Latin American Symposium on Circuits and Systems* (IEEE, 2016), pp. 215–218
59. Y. Narusue, Y. Kawahara, T. Asami, Maximizing the efficiency of wireless power transfer with a receiver-side switching voltage regulator. Wirel. Power Transf. 1–13 (2017). ISSN 2052-8418. https://doi.org/10.1017/wpt.2016.14
60. H. Li, J. Li, K. Wang, W. Chen, X. Yang, A maximum efficiency point tracking control scheme for wireless power transfer systems using magnetic resonant coupling. IEEE Trans. Power Electron. **30**(7), 3998–4008 (2015). ISSN 0885-8993. https://doi.org/10.1109/TPEL.2014.2349534
61. P. Nintanavongsa, U. Muncuk, D.R. Lewis, K.R. Chowdhury, Design optimization and implementation for RF energy harvesting circuits. IEEE J. Emerg. Sel. Top. Circuits Syst. **2**(1), 24–33 (2012). ISSN 2156-3357. https://doi.org/10.1109/JETCAS.2012.2187106
62. U. Guler, Y. Jia, M. Ghovanloo, A reconfigurable passive RF-to-DC converter for wireless IoT applications. IEEE Trans. Circuits Syst. II **66**(11), 1800–1804 (2019)
63. U. Guler, Y. Jia, M. Ghovanloo, A reconfigurable passive voltage multiplier for wireless mobile IoT applications. IEEE Trans. Circuits Syst. II **67**(4), 615–619 (2020)
64. H.-M. Lee, M. Ghovanloo, An integrated power-efficient active rectifier with offset-controlled high speed comparators for inductively powered applications. IEEE Trans. Circuits Syst. I **58** (8), 1749–1760 (2011)
65. L. Cheng, W.H. Ki, Y. Lu, T.S. Yim, Adaptive on/off delay-compensated active rectifiers for wireless power transfer systems. IEEE J. Solid-State Circuits **51**(3), 712–723 (2016). ISSN 0018-9200. https://doi.org/10.1109/JSSC.2016.2517119
66. C. Huang, T. Kawajiri, H. Ishikuro, A near-optimum 13.56 MHz CMOS active rectifier with circuit-delay real-time calibrations for high-current biomedical implants. IEEE J. Solid-State Circuits **51**(8), 1797–1809 (2016). ISSN 0018-9200. https://doi.org/10.1109/JSSC.2016.2582871
67. P. Pérez-Nicoli, F. Silveira, Comparator with self controlled delay for active rectifiers in inductive powering, in *IEEE Wireless Power Transfer Conference (WPTC)* (IEEE, 2018), pp. 1–3
68. X. Li, C.Y. Tsui, W.H. Ki, Power management analysis of inductively-powered implants with 1X/2X reconfigurable rectifier. IEEE Trans. Circuits Syst. I **62**(3), 617–624 (2015). ISSN 1549-8328. https://doi.org/10.1109/TCSI.2014.2366814
69. Q. Ma, M.R. Haider, Y. Massoud, A low-loss rectifier unit for inductive-powering of biomedical implants, in *IEEE/IFIP International Conference on VLSI-SoC* (2011), pp. 86–89. https://doi.org/10.1109/VLSISoC.2011.6081656
70. W.X. Zhong, S.Y.R. Hui, Maximum energy efficiency tracking for wireless power transfer systems. IEEE Trans. Power Electron. **30**(7), 4025–4034 (2015)
71. S. Stoecklin, T. Volk, A. Yousaf, L. Reindl, A maximum efficiency point tracking system for wireless powering of biomedical implants. Proc. Eng. **120**, 451–454 (2015). ISSN 1877-7058. doi: http://doi.org/10.1016/j.proeng.2015.08.666
72. X. Zhang, H. Lee, An efficiency-enhanced auto-reconfigurable 2×/3×SC charge pump for transcutaneous power transmission. IEEE J. Solid-State Circuits **45**(9), 1906–1922 (2010). ISSN 0018-9200. https://doi.org/10.1109/JSSC.2010.2055370
73. X. Zhang, H. Lee, A reconfigurable 2x/ 2.5x/ 3x/ 4x SC DC-DC regulator for enhancing area and power efficiencies in transcutaneous power transmissions, in *IEEE Custom Integrated Circuits Conference* (2011), pp. 1–4

Chapter 2
Inductive Link: Basic Theoretical Model

2.1 Reflected Load Theory in a 2-Coil Link

2.1.1 Underlying Physical Principles of Inductive Coupling: Self-Inductance (L), Mutual Inductance (M), and Coupling Coefficient (k)

Hans Christian Ørsted experimentally discovered in 1820 that electric currents create magnetic fields. A changing current thus creates a changing magnetic field. In 1831, Michael Faraday discovered that a changing magnetic field can induce electromotive force (i.e., voltage) across an electrical conductor. In essence, Faraday's law sets up the foundation that describes the physical principle behind the inductive WPT.

A change in the magnetic field, as a result of changing the current that passes through a conductor, can affect the same conductor that generates this field, and this effect, which appears in the form of a varying potential difference across the conductor, is particularly intense if the conductor is in the form of a loop. In fact a coil, which is a key building block of every inductive WPT system, can be considered a combination of multiple loops that are connected in series. This effect is modeled in circuit theory by defining a parameter, known as self-inductance (L). Assuming a sinusoidal excitation current i at frequency f, the amplitude of the induced sinusoidal voltage v would be $v = iL2\pi f$, which is a well-known relationship in circuit theory.

If a second conductor is exposed to the varying magnetic flux that is generated by the varying current in the first conductor, a varying voltage is induced across the second conductor as well. The induced voltage in the second conductor, produced by the current passing through the first conductor, is modeled by a parameter, known as mutual inductance (M). Assuming a sinusoidal excitation current (i) at frequency f passing through the first inductor, the induced voltage v across the

© The Author(s), under exclusive license to Springer Nature Switzerland AG 2021
P. Pérez-Nicoli et al., *Inductive Links for Wireless Power Transfer*,
https://doi.org/10.1007/978-3-030-65477-1_2

second conductor would be $v = iM2\pi f$. Based on superposition, the total voltage across each conductor (coil) in a pair that is in proximity of each other would then be the sum of their self-induced and mutually induced voltages with the sign of the latter defined based on the direction of their windings. This allows to work with a simple model constructed with lumped circuit components instead of a complex model based on the magnetic field.

In order to facilitate calculations and have a better understanding of the strength of the mutual coupling between two coils, this parameter is normalized by the self-inductance of the two inductors. The result would be a new parameter, known as the coupling coefficient, k, between the two coils, which is defined as:

$$k = M/\sqrt{L_1 L_2}, \tag{2.1}$$

where L_1 and L_2 are the self-inductance of each coil and M is the mutual inductance between them. This normalized coefficient, which varies between 0 and 1, represents how strongly the two coils are coupled, and it depends on the geometrical characteristics of the two coils, such as distance, radii, alignment, and orientation. When $k = 0$, the current in one coil does not affect the other one at all. On the other hand, when $k = 1$, it means that one coil affects the other as much as it affects itself. For instance, if we have two identical coils, $L_1 = L_2$, which are completely overlapping with zero distance between them, then $k \simeq 1$, which means that $L_1 = L_2 \simeq M$.

Once we understand the physical meaning of the coupling coefficient, k, we can foresee its strong impact on WPT systems and the fact that coils' geometry and orientation should be designed to provide the highest possible coupling coefficient. This is analytically proven in the efficiency expression deduced in this chapter.

2.1.2 Equivalent Circuit Model

Now we will consider the simplest model of a 2-coil WPT link in order to deduce the expressions for the key performance aspects: delivered power to the Rx-circuit, P_{MN}, and link efficiency, η_{Link}. Figure 2.1 shows this basic model of a 2-coil WPT system. In reality, each coil consists of its self-inductance, which is its desired characteristic defined above, and a few undesired characteristics that can be lumped into resistive and capacitive components. For the sake of simplicity, in this model, no parasitic capacitors are considered for the coils. R_{Tx} and R_{Rx} are the lumped parasitic resistances of L_{Tx} and L_{Rx}, respectively, which model the losses in their windings. The mutual inductance between the Tx and Rx coils is $M_{Tx\text{-}Rx}$, which depends on their geometry and relative position.

Note that the Tx-circuit and Rx-circuit blocks were defined, for the first time, in Sect. 1.3 (Fig. 1.3), and they include all the circuits connected before and after the Tx and Rx coils, respectively. In this section, compared with Fig. 1.3, we are substituting the Rx-circuit with its input impedance, Z_{MN}, to focus the analysis on

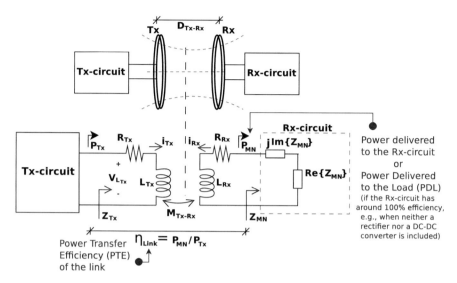

Fig. 2.1 A simplified schematic diagram of a 2-coil WPT link and its lumped circuit model

the inductive link. The subscript MN stands for matching network, which is the first block of the Rx-circuit as was depicted in Fig. 1.3. The deduction of Z_{MN}, considering that the Rx-circuit may include a rectifier and a DC-DC converter following the matching network, is addressed in Chap. 6. In Fig. 2.1, η_{Link} and P_{MN} are highlighted because they are the two main parameters that quantify the WPT link performance. These two parameters are deduced and widely used throughout the chapter.

If the matching networks of the Tx-circuit and the Rx-circuit are substituted by the typically used series or parallel resonators and assuming neither DC-DC converter nor rectifier are used, we obtain the four fundamental topologies presented in Fig. 2.2 [1].

From Fig. 2.2, comparing the series and parallel Rx resonators (Fig. 2.2a and b), it is evident that each case has a different Z_{MN}. In Chap. 5, we address the optimum value for Z_{MN} which determines the best Rx resonance topology, series or parallel, in a particular design. In this section, however, we only provide the following intuitive explanation. Note that the induced current in the Rx coil needs a return path, which in the case of a series resonator includes the load, R_L. If R_L is too large, with a series resonator, it is hard to induce a current in the Rx coil. At the limit, if R_L tends to infinity, no current is induced in the Rx series resonator; thus $P_{MN} = 0$ and $\eta_{Link} = 0$. In low-power applications such as AIMDs, where R_L is much greater than R_{Rx}, the parallel resonator provides the required return path avoiding the large R_L, thus achieving higher η_{Link} and P_{MN} than the series resonator. In high-power applications such as WPT for electric vehicles (EV), where R_L is comparable with R_{Rx}, higher η_{Link} and P_{MN} are obtained if all the induced current passes through the

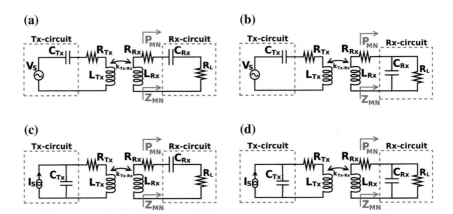

Fig. 2.2 The four fundamental topologies on the Tx and Rx side. In all cases, at resonance, we have $1/(C_{Tx}\omega) = \omega L_{Tx}$ and $1/(C_{Rx}\omega) = \omega L_{Rx}$. (**a**) Series-series, typically used with a voltage source Tx in high-power applications where R_L and R_{Rx} have the same order of magnitude. (**b**) Series-parallel, typically used with a voltage source Tx in low-power applications where R_L is much higher than R_{Rx}. (**c**) Parallel-series, typically used with a current Tx source in high-power applications where R_L and R_{Rx} have the same order of magnitude. (**d**) Parallel-parallel, typically used with a current Tx source in low-power applications where R_L is much higher than R_{Rx}

load; therefore, a series resonator is preferred. This intuitive analysis is quantified in an example in Sect. 2.1.6.

Regarding the Tx resonant structure, although both series and parallel topologies can be used, the voltage source with a series resonant capacitor is the most typically used topology. Therefore, the topologies of Fig. 2.2a and b are the most commonly used in various applications. In Chap. 7, the influence of the Tx type (voltage or current) and its resonant structure (series or parallel) in the closed-loop output voltage regulation are analyzed.

The PTE of the link, η_{Link}, for all the cases presented in Fig. 2.2 is calculated in Sect. 2.1.3, and the results are summarized in Table 2.1. The P_{MN}, also for all the cases, is calculated in Sect. 2.1.4 and summarized in Table 2.3.

Although some authors analyze each topology of Fig. 2.2 separately [2], in this chapter, we present a general approach that applies to all variations. The analysis is carried out based on Fig. 2.1, which includes not only the four particular cases of Fig. 2.2 but also other more sophisticated resonant structures as the series-parallel Rx proposed in [3, 4]. These other resonant structures will be introduced and analyzed in Chap. 5, Table 5.4.

2.1.3 Calculation of Link Efficiency, η_{Link}

Since there is no radiation in the near-field regime, all the energy losses of the link occur in the coils' parasitic resistances, R_{Tx} and R_{Rx}, shown in Fig. 2.1. Therefore, the power provided by the Tx-circuit, P_{Tx}, is either delivered to the Rx-circuit, P_{MN}, or dissipated as heat in the coil resistors. The Power Transfer Efficiency (PTE) of the link, η_{Link}, defined in Fig. 1.3, is given by the ratio between P_{MN} and P_{Tx}, Fig. 2.1. To calculate η_{Link}, the efficiency in each coil is calculated as follows.

The efficiency of the Rx coil is defined as the ratio between the power delivered to Z_{MN}, P_{MN}, and the total power dissipated in R_{Rx} and Z_{MN}:

$$\eta_{L_{Rx}} = \frac{Re\{Z_{MN}\}i_{Rx}^2/2}{Re\{Z_{MN}\}i_{Rx}^2/2 + R_{Rx}i_{Rx}^2/2}$$
$$= \frac{1/R_{Rx}}{1/R_{Rx} + 1/Re\{Z_{MN}\}} \frac{R_{Rx} Re\{Z_{MN}\}}{R_{Rx} Re\{Z_{MN}\}} \tag{2.2}$$

where i_{Rx} is the Rx peak current and $Re\{Z_{MN}\}$ indicates the real part of Z_{MN}. Multiplying numerator and denominator of (2.2) by ωL_{Rx}, the result can be expressed as in (2.5), in terms of the Rx coil quality factor:

$$Q_{Rx} = \frac{\omega L_{Rx}}{R_{Rx}}, \tag{2.3}$$

and load quality factor:

$$Q_L = \frac{\omega L_{Rx}}{Re\{Z_{MN}\}}. \tag{2.4}$$

$$\eta_{L_{Rx}} = \frac{Q_{Rx}}{Q_{Rx} + Q_L} \tag{2.5}$$

To determine the efficiency of the Tx coil, Z_{Tx} (see Fig. 2.1) needs to be computed. Z_{Tx} includes the effect of the Rx through the mutual inductance:

$$\left. \begin{array}{l} \text{Kirchhoff's laws} \\ \text{in Fig. 2.1} \end{array} \Rightarrow \begin{array}{l} V_{L_{Tx}} = (R_{Tx} + j\omega L_{Tx})i_{Tx} + j\omega M_{Tx\text{-}Rx}i_{Rx} \\ i_{Rx} = \frac{-i_{Tx}j\omega M_{Tx\text{-}Rx}}{R_{Rx} + j\omega L_{Rx} + Z_{MN}} \end{array} \right\} \Rightarrow$$
$$\Longrightarrow Z_{Tx} = \frac{V_{L_{Tx}}}{i_{Tx}} = (R_{Tx} + j\omega L_{Tx}) + \underbrace{\frac{\omega^2 M_{Tx\text{-}Rx}^2}{R_{Rx} + j\omega L_{Rx} + Z_{MN}}}_{Z_{Rx\text{-}Tx_{ref}}} \tag{2.6}$$

As can be seen in Fig. 2.1 and (2.6), Z_{Tx} is the impedance seen into the Tx coil, which is a series combination of R_{Tx} and L_{Tx} and a reflected impedance from the Rx coil, $Z_{Rx\text{-}Tx_{ref}}$, defined in (2.6).

The Tx coil efficiency is the power delivered to $Re\{Z_{Rx\text{-}Tx_{ref}}\}$ (power transferred to the Rx coil), divided by the total power dissipated in R_{Tx} and $Re\{Z_{Rx\text{-}Tx_{ref}}\}$:

$$\eta_{L_{Tx}} = \frac{Re\{Z_{Rx\text{-}Tx_{ref}}\}i_{Tx}^{2}/2}{Re\{Z_{Rx\text{-}Tx_{ref}}\}i_{Tx}^{2}/2 + R_{Tx}i_{Tx}^{2}/2}, \qquad (2.7)$$

where i_{Tx} is the Tx peak current.

The maximum Tx coil efficiency is obtained when the real part of this reflected impedance is maximized. It occurs when the imaginary part of $j\omega L_{Rx} + Z_{MN}$ is equal to zero (resonant Rx). In this case, the reflected impedance is maximum, and it is purely real, which is the advantage of having a resonant Rx coil. This is the main reason why a Rx matching network is included in the Rx-circuit, as discussed in Sect. 1.3.5, to achieve $\omega L_{Rx} = -Im\{Z_{MN}\}$. In Sect. 2.1.6, we will quantify, in an example, the effect of the Rx resonance in η_{Link}.

In the case of a resonant Rx coil, an expression for this reflected resistance can be derived from (2.6) as follows:

$$\left.\begin{array}{c} \overset{(2.6)}{\text{resonant Rx coil}} \Rightarrow \omega L_{Rx} = -Im\{Z_{MN}\} \\ (2.1) \Rightarrow M_{Tx\text{-}Rx} = k_{Tx\text{-}Rx}\sqrt{L_{Tx}L_{Rx}} \end{array}\right\} \Rightarrow$$

$$\Rightarrow Z_{Rx\text{-}Tx_{ref}} = \frac{\omega^{2}\overbrace{k_{Tx\text{-}Rx}^{2}L_{Tx}L_{Rx}}^{M_{Tx\text{-}Rx}^{2}}}{R_{Rx} + j\omega L_{Rx} + jIm\{Z_{MN}\} + Re\{Z_{MN}\}} \Rightarrow \qquad (2.8)$$

$$\Rightarrow Z_{Rx\text{-}Tx_{ref}} = Re\{Z_{Rx\text{-}Tx_{ref}}\} = R_{Rx\text{-}Tx_{ref}} = \frac{\omega^{2}k_{Tx\text{-}Rx}^{2}L_{Tx}L_{Rx}}{R_{Rx} + Re\{Z_{MN}\}}.$$

Then, using the definitions of Rx coil quality factor, Q_{Rx} (2.3), and load quality factor, Q_L (2.4), and defining the Tx coil quality factor as

$$Q_{Tx} = \frac{\omega L_{Tx}}{R_{Tx}}, \qquad (2.9)$$

the $R_{Rx\text{-}Tx_{ref}}$ (2.8) can be rewritten in terms of the quality factors as presented in (2.10):

$$\overset{\substack{(2.3)\\(2.4)\\(2.9)}}{(2.8) \Longrightarrow} Re\{Z_{Rx\text{-}Tx_{ref}}\} = R_{Rx\text{-}Tx_{ref}} = k_{Tx\text{-}Rx}^{2}Q_{Tx}\overbrace{\frac{Q_{Rx}Q_L}{Q_L + Q_{Rx}}}^{Q_{Rx\text{-}L}}R_{Tx}$$

$$= k_{Tx\text{-}Rx}^{2}Q_{Tx}Q_{Rx\text{-}L}R_{Tx}, \qquad (2.10)$$

where $Q_{Rx\text{-}L}$ was defined as:

$$Q_{Rx\text{-}L} = \frac{Q_{Rx}Q_L}{Q_{Rx} + Q_L} = \frac{\omega L_{Rx}}{R_{Rx} + Re\{Z_{MN}\}}. \tag{2.11}$$

The equivalent circuit, with the mutual coupling modeled by the reflected impedance, assuming a resonant Rx, $\omega L_{Rx} = -Im\{Z_{MN}\}$, is presented in Fig. 2.3. The resulting Tx coil efficiency can be rewritten from (2.7) and (2.10) as:

$$\eta_{L_{Tx}} = \frac{R_{Rx\text{-}Tx_{ref}}}{R_{Rx\text{-}Tx_{ref}} + R_{Tx}} = \frac{k_{Tx\text{-}Rx}{}^2 Q_{Tx} Q_{Rx\text{-}L}}{k_{Tx\text{-}Rx}{}^2 Q_{Tx} Q_{Rx\text{-}L} + 1}. \tag{2.12}$$

Finally, the total link efficiency from (2.5) and (2.12) ($\eta_{Link} = \eta_{L_{Tx}} . \eta_{L_{Rx}}$) will be:

$$\eta_{Link} = \underbrace{\frac{Q_{Rx\text{-}L}}{Q_L}}_{\eta_{L_{Rx}}} \underbrace{\frac{k_{Tx\text{-}Rx}{}^2 Q_{Tx} Q_{Rx\text{-}L}}{k_{Tx\text{-}Rx}{}^2 Q_{Tx} Q_{Rx\text{-}L} + 1}}_{\eta_{L_{Tx}}}, \tag{2.13}$$

where the Rx coil efficiency (2.5) was rewritten using the previously defined $Q_{Rx\text{-}L}$ (2.11), $\eta_{L_{Rx}} = \frac{Q_{Rx}}{Q_{Rx}+Q_L} = \frac{Q_{Rx\text{-}L}}{Q_L}$.

As presented in Chap. 1, Fig. 1.3, the first block of the Rx-circuit is a matching network which adapts its input impedance, Z_{MN}. Equation 2.13 is valid for any matching network that holds resonance ($\omega L_{Rx} = -Im\{Z_{MN}\}$) as it was the only assumption about it. However, the load quality factor, Q_L, depends on $Re\{Z_{MN}\}$ (2.4), which is affected by the Rx matching network. Table 2.1 presents the value of Q_L in the typical cases of parallel and series resonators in the fundamental topologies of Fig. 2.2. In both cases, η_{Link} is calculated using (2.13) where Q_L is obtained from (2.14) for the Rx parallel resonator and from (2.15) for the Rx series resonator.

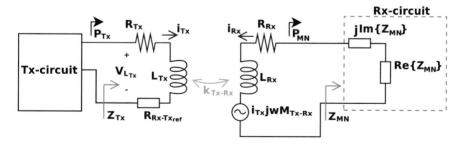

Fig. 2.3 The 2-coil link model assuming a resonant Rx, $\omega L_{Rx} = -Im\{Z_{MN}\}$. The coupling coefficient is in gray as it is modeled by the reflected resistance on the Tx side and the induced voltage on the Rx side

Table 2.1 Q_L and η_{Link} calculation for parallel and series resonators in the Rx. Note that these expressions are valid for any Tx-circuit topology

Rx Parallel-resonator	Rx Series-resonator
$Z_{MN} = \dfrac{R_L}{1+j\omega C_{Rx}R_L} = \dfrac{R_L(1-j\omega C_{Rx}R_L)}{1+(\omega C_{Rx}R_L)^2}$ assuming $(\omega C_{Rx}R_L)^2 \gg 1$ ←(see footnote 1) $\left.\vphantom{\rule{0pt}{3em}}\right\}$ $\Rightarrow Z_{MN} = \dfrac{1}{R_L(\omega C_{Rx})^2} - \dfrac{j}{\omega C_{Rx}}$	$Z_{MN} = R_L - \dfrac{j}{\omega C_{Rx}}$
$Q_L = \dfrac{\omega L_{Rx}}{Re\{Z_{MN}\}} = \dfrac{\omega L_{Rx}}{1/(R_L(\omega C_{Rx})^2)}$ assuming resonant Rx, $\omega C_{Rx} = \frac{1}{\omega L_{Rx}}$ $\left.\vphantom{\rule{0pt}{3em}}\right\}$ (2.14) $\Rightarrow Q_L = \dfrac{R_L}{\omega L_{Rx}}$	$Q_L = \dfrac{\omega L_{Rx}}{R_L}$ (2.15)

For both, parallel and series resonators,

$$\eta_{Link} = \frac{Q_{Rx\text{-}L}}{Q_L}\frac{k_{Tx\text{-}Rx}{}^2 Q_{Tx}Q_{Rx\text{-}L}}{k_{Tx\text{-}Rx}{}^2 Q_{Tx}Q_{Rx\text{-}L}+1} \tag{2.13}$$

$$\text{where,} \qquad Q_{Rx\text{-}L} = \frac{Q_{Rx}Q_L}{Q_{Rx}+Q_L} \tag{2.11}$$

Kiani et al. in [5, 6], with a parallel resonator, defined Q_L as in (2.14), while in [7], with a series resonator, Q_L was defined as in (2.15). The series and parallel Rx resonators are further discussed in [8] where Q_L for each case is calculated. In this book, we use the more general definition of Q_L in (2.4), which is valid not only for series and parallel resonators but also for any matching network that satisfies the Rx resonance condition ($\omega L_{Rx} = -Im\{Z_{MN}\}$), as the ones introduced in [3, 4].

In Table 2.2 we present three numerical examples were η_{Link} is calculated using (2.13) along with their associated references, where more details about these links can be found.

So far in this section, we have addressed the advantages of having a resonant Rx, highlighting that a resonant Rx maximizes the real part of the reflected impedance, $Re\{Z_{Rx\text{-}Tx_{ref}}\}$, increasing the link efficiency, η_{Link}. In fact, as mentioned, the

---

[1]$(\omega C_{Rx}R_L)^2 \gg 1$ is assumed because otherwise the imaginary part of Z_{MN} will depend on the load, R_L. If that is the case, a modification in the load circuit power consumption may affect the Rx resonance which is not desired. Therefore, when a parallel resonant capacitor is used, this condition should be satisfied.

Table 2.2 Numerical examples

	Application	f	Q_{Tx}	Q_{Rx}	Q_L	$Q_{Rx\text{-}L}$	$k_{Tx\text{-}Rx}$	η_{Link}
[9]	EV	20 kHz	935	760	4.4	4.3	0.15	98%
[10]	RFID	134.2 kHz	23.5	47	524	43.2	0.02	2.6%
[11]	AIMD	6.78 MHz	153	32	9	7	0.04	49%

deduced expression for η_{Link} (2.13) assumes that the Rx matching network is holding the Rx resonance ($\omega L_{Rx} = -Im\{Z_{MN}\}$). However, whether the Tx was at resonance or not was not taken into account. This is because based on the definition of η_{Link} in Fig. 2.1, it is not necessary to consider the Tx-circuit which drives the Tx coil. Therefore, the η_{Link} (2.13) does not depend on the Tx resonance. Nevertheless, the Tx resonance does affect the Tx-circuit performance. With a non-resonant Tx, a higher voltage is required to generate a certain amount of current in both Tx and Rx loops. Additionally, the efficiency of the inverter depends on its load, as was mentioned in Sect. 1.3.2. This is the reason why a Tx matching network is included in the Tx-circuit as was discussed in Sect. 1.3.3.

2.1.4   Calculation of Power Delivered to the Rx-circuit, P_{MN}

The power delivered to the Rx-circuit, P_{MN}, can be calculated from η_{Link} as:

$$P_{MN} = \eta_{Link} P_{Tx}. \tag{2.16}$$

Although the η_{Link} does not depend on the Tx-circuit, as discussed in the previous section, P_{Tx} is determined by the Tx-circuit. In this section, we consider the four cases depicted in Table 2.3 for the Tx-circuit. In the four cases, the source impedance has been ignored for the sake of simplicity. Table 2.3 summarizes the results for the P_{MN} in each case, and its associated analysis is presented in Appendixes A.1, A.2, A.3, and A.4. In all the cases, Q_L is the one defined in (2.4), and both Tx and Rx are assumed to be at resonance ($\omega L_{Tx} = 1/(\omega C_{Tx})$ and $\omega L_{Rx} = -Im\{Z_{MN}\}$) (Figs. 2.4, 2.5, 2.6, and 2.7).

2.1.5   Effects of Coils' Quality Factor (Q) and Coupling Coefficient (k) on the Link

In this section, we take a closer look at the effects of Q and k on the link characteristics. The quality factor of a coil is the ratio between its reactance ωL and its parasitic resistance R, $Q = \omega L/R$. Therefore, a $Q < 1$ means that the resistive (dissipative) effect dominates over the inductive effect. By looking at the

Table 2.3 P_{MN} for different Tx-circuits. Expressions (2.17), (2.18), (2.19), and (2.20) are valid for any resonant Rx-circuit and a resonant Tx ($1/(C_{Tx}\omega) = \omega L_{Tx}$)

	See mathematical proof in Appendix A.1
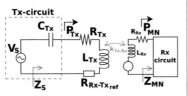 **Fig. 2.4** Voltage source and series resonant Tx	$$P_{MN} = \frac{V_S^2}{2R_{Tx}} \frac{Q_{Rx\text{-}L}}{Q_L} \frac{k_{Tx\text{-}Rx}{}^2 Q_{Tx} Q_{Rx\text{-}L}}{(k_{Tx\text{-}Rx}{}^2 Q_{Tx} Q_{Rx\text{-}L} + 1)^2}$$ (2.17)
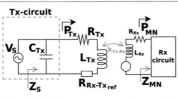 **Fig. 2.5** Voltage source and parallel resonant Tx	See mathematical proof in Appendix A.2 $$P_{MN} = \frac{V_S^2}{2(\omega L_{Tx})^2} R_{Tx} \frac{Q_{Rx\text{-}L}}{Q_L} k_{Tx\text{-}Rx}{}^2 Q_{Tx} Q_{Rx\text{-}L}$$ (2.18)
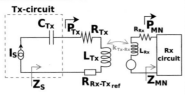 **Fig. 2.6** Current source and series resonant Tx	See mathematical proof in Appendix A.3 $$P_{MN} = \frac{I_S^2}{2} R_{Tx} \frac{Q_{Rx\text{-}L}}{Q_L} k_{Tx\text{-}Rx}{}^2 Q_{Tx} Q_{Rx\text{-}L}$$ (2.19)
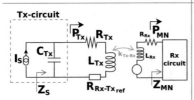 **Fig. 2.7** Current source and parallel resonant Tx	See mathematical proof in Appendix A.4 $$P_{MN} = \frac{I_S^2(\omega L_{Tx})^2}{2R_{Tx}} \frac{Q_{Rx\text{-}L}}{Q_L} \cdot$$ $$\cdot \frac{k_{Tx\text{-}Rx}{}^2 Q_{Tx} Q_{Rx\text{-}L}}{(k_{Tx\text{-}Rx}{}^2 Q_{Tx} Q_{Rx\text{-}L} + 1)^2}$$ (2.20)

For all the cases: $Q_L = \dfrac{\omega L_{Rx}}{Re\{Z_{MN}\}}$ (2.4) $Q_{Rx\text{-}L} = \dfrac{Q_{Rx} Q_L}{Q_{Rx} + Q_L}$ (2.11)

Q definition, we can conclude that if we want to generate a magnetic field with low losses, the higher Q, the better.

The coupling coefficient was introduced in Sect. 2.1.1, and it is a number between zero and one, which represents how strongly two coils are coupled.

This previous discussion is consistent with the equation derived for η_{Link} (2.13), which suggests that if $k_{Tx\text{-}Rx}$, Q_{Tx}, or Q_{Rx} raise, η_{Link} (2.13) increases. This is one of the fundamentals of inductive powering. High-quality factors and coupling coefficient are desired to reduce losses and increase η_{Link} under almost any condition. However, there are other factors that often limit the maximum values of these key parameters in practical applications. In Fig. 2.8, an example is presented to quantify this dependency for an inductive link with a $Q_L = 50$. In Chap. 3, practical aspects of the coil design are addressed, providing a better understanding of what level of k and Q can be achieved in a practical WPT design for a particular application. Depending on the coils' geometry and operating frequency, which in turn depend on the WPT application, we can anticipate that a Q of 10,000 could be difficult to achieve, while a Q of 1000 is possible in WPT for electric vehicles, a Q of 200 is possible in non-miniaturized AIMD applications, and a Q of 10–20 is achieved with integrated on-chip coils. Regarding the coupling coefficient, a $k_{Tx\text{-}Rx} = 0.5$ can be achieved, for instance, with two identical circular coils with a diameter of 20 cm, separated by 8 cm. With the same coils, $k_{Tx\text{-}Rx}$ drops to only 0.01 and 0.005 when the distance between the coils is increased to 45 and 60 cm,

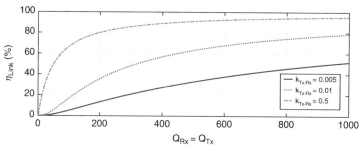

The η_{Link} is plotted using:

$$\eta_{Link} = \frac{Q_{Rx\text{-}L}}{Q_L}\,\frac{k_{Tx\text{-}Rx}{}^2 Q_{Tx} Q_{Rx\text{-}L}}{k_{Tx\text{-}Rx}{}^2 Q_{Tx} Q_{Rx\text{-}L} + 1} \qquad\qquad Q_{Rx\text{-}L} = \frac{Q_{Rx} Q_L}{Q_{Rx} + Q_L}$$
$$(2.13) \qquad\qquad\qquad\qquad\qquad\qquad\qquad\qquad\qquad (2.11)$$

- $Q_{Tx} = Q_{Rx}$ ← Horizontal axis
- $k_{Tx\text{-}Rx}$ ← The value is indicated in the plot legend

- $Q_L = 50$, selected for the proof-of-concept system. The design of Q_L is addressed in Chapter 5.

Fig. 2.8 Example to quantify the dependency of η_{Link} (2.13), with the coupling coefficient and quality factor. The calculation script is available in the supplementary material, file Sec215

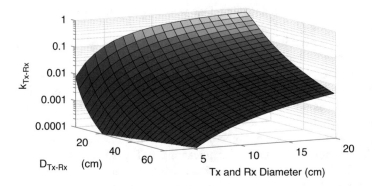

Fig. 2.9 Coupling coefficient, $k_{Tx\text{-}Rx}$, for two identical coils as a function of coils' diameter and separation. $k_{Tx\text{-}Rx}$ was approximated using (3.16) which is introduced in Chap. 3. The calculation script is available in the supplementary material, file Sec215

Table 2.4 Electromagnetic field solvers

Tool	Provider
CST studio Suite	SIMULIA
Sim4Life	Zurich MedTech
HFSS	ANSYS
Multiphysics	COMSOL
CLS	ITIS Foundation
FEKO	Altair Engineering

respectively. If the separation is kept at 8 cm, $k_{Tx\text{-}Rx}$ drops to 0.01 and 0.005 when both coils' diameters decrease to 4 and 3 cm, respectively. The coupling coefficient between two identical coils as a function of coils' diameter and separation is plotted in Fig. 2.9.

While Q can be easily calculated from the electrical characteristics of a coil, the self-inductance L, the series parasitic resistance R, and the coupling coefficient k are highly dependent on the coil's geometry and the complex distribution of electromagnetic flux around one coil or in between two coupled coils. Therefore, while researchers have come up with certain models for a few simpler and more popular geometries, such as a circular conductive loop [12, 13] or squared-shaped Printed Spiral Coils (PSC) [14–21], the most accurate method to derive these parameters is Finite Element Analysis (FEA). The FEA is often implemented in the form of software that is specifically developed for this purpose, which is generally referred to as electromagnetic field solver. A few popular field solvers and their providers are listed in Table 2.4. In Chap. 3, the theoretical calculation and simulation of self-inductance and parasitic resistance among other electrical characteristics of the coil are discussed in more details.

The effects of $k_{Tx\text{-}Rx}$, Q_{Tx}, and Q_{Rx} on the power delivered to the Rx-circuit, P_{MN}, are more complex than their effects on the PTE of the link, η_{Link}. The Tx-circuit output power, P_{Tx}, depends on its load impedance, which includes the

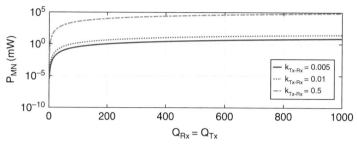

The P_{MN} is plotted using:

$$P_{MN} = \frac{I_S^2}{2} R_{Tx} \frac{Q_{Rx\text{-}L}}{Q_L} k_{Tx\text{-}Rx}{}^2 Q_{Tx} Q_{Rx\text{-}L} \qquad (2.19) \qquad\qquad Q_{Rx\text{-}L} = \frac{Q_{Rx} Q_L}{Q_{Rx} + Q_L} \qquad (2.11)$$

- $Q_{Tx} = Q_{Rx} \leftarrow$ Horizontal axis
- $k_{Tx\text{-}Rx} \leftarrow$ The value is indicated in the plot legend

- $Q_L = 50$; • $I_S = 50$ mA • $R_{Tx} = 5\,\Omega$

Fig. 2.10 Example to quantify the dependency of P_{MN} (2.19), with the coupling coefficient $k_{Tx\text{-}Rx}$ and quality factor. This example is for a current Tx source with a series resonant capacitor Fig. 2.6. The calculation script is available in the supplementary material, file Sec215

reflected resistance $R_{Rx-Tx_{ref}}$ (2.10), which in turn depends on $k_{Tx\text{-}Rx}$, Q_{Tx}, and Q_{Rx}. Therefore, the effect of these parameters on $P_{MN} = \eta_{Link} P_{Tx}$ is not trivial as it not only affects the efficiency, η_{Link}, but also P_{Tx}. In the case of a current source with a series resonant capacitor on the Tx side, as shown in Fig. 2.6, an increase in $k_{Tx\text{-}Rx}$, Q_{Tx}, or Q_{Rx} increases both η_{Link} and P_{Tx}. This case is depicted in Fig. 2.10. However, with a voltage source and a series resonant capacitor on the Tx side (Fig. 2.4), an increase in $k_{Tx\text{-}Rx}$, Q_{Tx}, or Q_{Rx} increases the reflected resistance (2.10) reducing P_{Tx}. The P_{MN} for this latter case is presented in Fig. 2.11. Although in this case P_{MN} could be reduced due to an increase in $k_{Tx\text{-}Rx}$, Q_{Tx}, or Q_{Rx} it does not mean a performance degradation. This just indicates that a higher Tx voltage, V_S, is required because the load resistance of the Tx-circuit is higher.

2.1.6   Effect of Tx and Rx Resonance on the Link

In this section, the effect of the Tx and Rx resonance on η_{Link} and P_{MN} is quantified. To do that, we calculate η_{Link} and P_{MN} for the four exemplar systems presented in Fig. 2.12 and Table 2.5, where all combinations of resonance or non-resonance are considered for Tx and Rx. In the deduction of η_{Link} in (2.13), we assume to have a resonant Rx, and the expressions for P_{MN} summarized in Table 2.3 assumes resonance in both Rx and Tx. Therefore, in Table 2.5, the deduction of the analytical

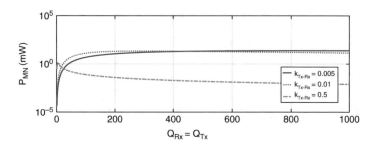

The P_{MN} is plotted using:

$$P_{MN} = \frac{V_S^2}{2R_{Tx}} \frac{Q_{Rx\text{-}L}}{Q_L} \frac{k_{Tx\text{-}Rx}^2 Q_{Tx} Q_{Rx\text{-}L}}{(k_{Tx\text{-}Rx}^2 Q_{Tx} Q_{Rx\text{-}L} + 1)^2} \qquad (2.17) \qquad\qquad Q_{Rx\text{-}L} = \frac{Q_{Rx} Q_L}{Q_{Rx} + Q_L} \qquad (2.11)$$

- $Q_{Tx} = Q_{Rx} \leftarrow$ Horizontal axis
- $k_{Tx\text{-}Rx} \leftarrow$ The value is indicated in the plot legend

- $Q_L = 50$; • $V_S = 1$ V • $R_{Tx} = 5$ Ω

Fig. 2.11 Example to quantify the dependency of P_{MN} (2.17), with the coupling coefficient $k_{Tx\text{-}Rx}$ and quality factor. This example is for a voltage source Tx with a series resonant capacitor (Fig. 2.4). The calculation script is available in the supplementary material, file Sec215

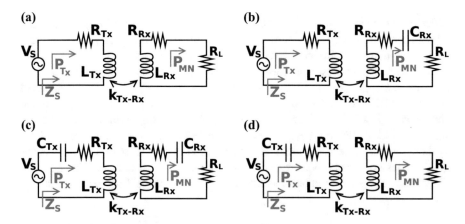

Fig. 2.12 Cases analyzed in Table 2.5. (**a**) Case A: Non-resonant Tx, non-resonant Rx. (**b**) Case B: Non-resonant Tx, resonant Rx. (**c**) Case C: Resonant Tx, resonant Rx. (**d**) Case D: Resonant Tx, non-resonant Rx

expressions is presented in order to consider the non-resonant condition that is addressed in this section.

From the results of the examples shown in Table 2.5, it can be seen that when the Rx is not resonating, cases A and D, the reflected impedance, $Z_{Rx\text{-}Tx_{ref}}$, is mostly reactive, and its real part is much smaller than the case where the Rx is resonating,

Table 2.5 Quantification of Tx and Rx resonance. The calculation script is available in the supplementary material, file Sec216

Parameters for all the table
$R_{Tx} = 5\,\Omega$; $Q_{Tx} = Q_{Rx} = 200$; $Q_L = 50$; $k_{Tx\text{-}Rx} = 0.01$; $V_S = 1\,\text{V}$

Case A: Fig. 2.12a, Non-resonant Tx, non-resonant Rx

$$(2.6) \Rightarrow Z_{Rx\text{-}Tx_{ref}} = \frac{\omega^2 M_{Tx\text{-}Rx}{}^2}{R_{Rx} + j\omega L_{Rx} + R_L}$$

$$= \frac{k_{Tx\text{-}Rx}{}^2 Q_{Tx} Q_{Rx}}{1 + Q_{Rx}(j + 1/Q_L)} R_{Tx} \quad (2.21)$$

$$Z_{Rx\text{-}Tx_{ref}} = (2.5 - j100)\,\text{m}\Omega$$

$$\left.\begin{array}{l} \eta_{Link} = \eta_{L_{Tx}} \cdot \eta_{L_{Rx}} \\ (2.5) \\ (2.7) \end{array}\right\} \Longrightarrow \qquad (2.22) \qquad \eta_{Link} = 0.04\%$$

$$\eta_{Link} = \frac{Q_{Rx}}{Q_{Rx} + Q_L} \frac{Re\{Z_{Rx\text{-}Tx_{ref}}\}}{Re\{Z_{Rx\text{-}Tx_{ref}}\} + R_{Tx}}$$

$$P_{MN} = \overbrace{\frac{V_S^2}{2|Z_S|^2} Re\{Z_S\}}^{P_{Tx}} \eta_{Link}$$

$$P_{MN} = 1\,\text{nW}$$

where: $Z_S = R_{Tx} + j\omega L_{Tx} + Z_{Rx\text{-}Tx_{ref}}$

$$= R_{Tx}\left(1 + jQ_{Tx} + \frac{Z_{Rx\text{-}Tx_{ref}}}{R_{Tx}}\right) \quad (2.23)$$

Case B: Fig. 2.12b, Non-resonant Tx, resonant Rx

$Z_{Rx\text{-}Tx_{ref}}$ same deduced in (2.8)	$Z_{Rx\text{-}Tx_{ref}} = 4\,\Omega$
η_{Link} same deduced in (2.13)	$\eta_{Link} = 36\%$
P_{MN} same deduced in case A (2.23)	$P_{MN} = 1.6\,\mu\text{W}$

Case C: Fig. 2.12c, Resonant Tx, resonant Rx

$Z_{Rx\text{-}Tx_{ref}}$ same deduced in (2.8)	$Z_{Rx\text{-}Tx_{ref}} = 4\,\Omega$
η_{Link} same deduced in (2.13)	$\eta_{Link} = 36\%$
P_{MN} same deduced in (2.17)	$P_{MN} = 20\,\text{mW}$

Case D: Fig. 2.12d, Resonant Tx, non-resonant Rx

$Z_{Rx\text{-}Tx_{ref}}$ same deduced in case A (2.21)	$Z_{Rx\text{-}Tx_{ref}} = (2.5 - j100)\,\text{m}\Omega$
η_{Link} same deduced in case A (2.22)	$\eta_{Link} = 0.04\%$

$$P_{MN} = \overbrace{\frac{V_S^2}{2|Z_S|^2} Re\{Z_S\}}^{P_{Tx}} \eta_{Link} \qquad (2.24)$$

$$P_{MN} = 40\,\mu\text{W}$$

where: $Z_S = R_{Tx} + Z_{Rx\text{-}Tx_{ref}}$

cases B and C. Therefore, from (2.22), it can be seen that low $Re\{Z_{Rx\text{-}Tx_{ref}}\}$ deteriorates η_{Link} and thus P_{MN} in cases A and D. On the other hand, with a resonant Rx, the reflected impedance is real and greater than the non-resonant case, increasing η_{Link} and thus the P_{MN}.

As discussed in Sect. 2.1.3, the Tx resonance does not affect η_{Link}. This is due to the η_{Link} definition, $\eta_{Link} = P_{MN}/P_{Tx}$, and the statement is true even if the Tx source output impedance is considered. When the Tx source output impedance is present, which is the case in any actual implementation, the resonance or lack thereof of the Tx coil only affects the Tx-circuit efficiency, η_{Tx}, defined in Fig. 1.3. In the cases considered in Table 2.5, an output resistance of the Tx voltage source could be considered as being included in R_{Tx}, and therefore, all the equations presented there are still valid.

For a given Tx voltage source, regardless of whether or not its output impedance is considered, P_{MN} is dramatically increased by adding a series resonant capacitor. This is because the resonant capacitor decreases the modulus of the impedance that loads the Tx voltage source, $|Z_S|$, increasing its output current, $V_S/|Z_S|$, and its output power $P_{Tx} = \frac{1}{2}\frac{V_S^2}{|Z_S|^2}Re\{Z_S\}$, which improves the power delivered to the Rx-circuit, $P_{MN} = P_{Tx}\eta_{Link}$.

In this section, we also compare the series versus parallel Rx resonator using the same exemplar system. In Table 2.6, η_{Link} and P_{MN} for the typical series-series and series-parallel typologies are plotted (Figs. 2.13 and 2.14). As was anticipated in Sect. 2.1.2, the series resonator has a better performance for low R_L (high-power applications), while the parallel resonator performs better for a larger R_L (low-power applications).

2.1.7   Frequency Splitting Effect

Comparing Figs. 2.10 and 2.11, we observe a rather strange phenomenon in the latter. Generally we expect P_{MN} to increase with higher $k_{Tx\text{-}Rx}$, which is what we observe in Fig. 2.10. However, in Fig. 2.11, above a certain level, when both $k_{Tx\text{-}Rx}$ and $Q_{Tx} = Q_{Rx}$ are high, a significant drop is observed in P_{MN}. What is observed here is that when a voltage source is used on the Tx side, large $k_{Tx\text{-}Rx}$, Q_{Tx}, and/or Q_{Rx} reduce P_{MN} due to the reflected impedance being too large, which reduces the Tx-circuit output power, P_{Tx}.

Changing the carrier frequency reduces the reflected impedance, $Z_{Rx\text{-}Tx_{ref}}$ (2.6), resulting in increased P_{Tx} and consequently P_{MN}. This effect of achieving higher P_{MN} by operating at a frequency that is different from the resonance frequency of each individual coil ($\omega_{res} = 1/\sqrt{L_{Tx}C_{Tx}} = 1/\sqrt{L_{Rx}C_{Rx}}$) is known as the frequency splitting effect.

Another way to address this frequency splitting phenomenon is considering the two resonant coils as two coupled resonators [22, 23]. Two coupled L-C (inductor-capacitor) oscillators have two natural frequencies. If we consider two identical

Table 2.6 Comparison between series and parallel Rx resonators. The calculation script is available in the supplementary material, file Sec216

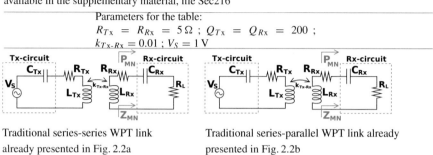

Parameters for the table:
$$R_{Tx} = R_{Rx} = 5\,\Omega \; ; \; Q_{Tx} = Q_{Rx} = 200 \; ;$$
$$k_{Tx\text{-}Rx} = 0.01 \; ; \; V_S = 1 \text{ V}$$

Traditional series-series WPT link already presented in Fig. 2.2a | Traditional series-parallel WPT link already presented in Fig. 2.2b

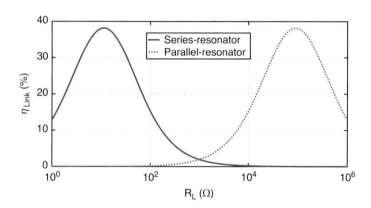

Fig. 2.13 η_{Link} calculated using (2.13)

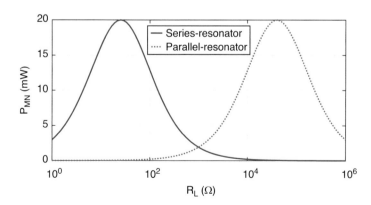

Fig. 2.14 P_{MN} calculated using (2.17)

L-C resonators, with a mutual inductance M, the natural frequencies of the system are $\omega+ = 1/\sqrt{(L + M)C}$ and $\omega- = 1/\sqrt{(L - M)C}$. When M is much lower than L, both natural frequencies are equal to the resonance frequency of each individual oscillator $(1/\sqrt{LC})$. However, when M is not negligible compared to L, the two different natural frequencies are observed. If we drive the circuit, the largest amplitude of oscillation occurs when the driving frequency is close to one of the natural frequencies [22]. This means that the highest P_{MN}, which is a function of the amplitude of oscillation, is obtained when the system is forced/driven at one of the natural frequencies. In [6], Kiani et al. calculated η_{Link} and P_{MN} considering the two resonant coils as two coupled resonators [23] (coupled-mode theory). Additionally, they compared the results with the ones obtained from a circuit perspective (perspective of this book), proving that despite using different parameters and terminologies, both approaches lead to the exact same set of equations.

For the sake of simplicity, consider the simple series-series resonant WPT system shown in Fig. 2.15. Let us assume that the resonance capacitors were selected to resonate at 13.56 MHz, $Q_{Tx} = Q_{Rx} = 200$, $Q_L = 50$, $R_{Tx} = 5\,\Omega$, and $V_S = 1$ V. In order to observe the frequency splitting effect, the P_{MN} for this example system is presented in Fig. 2.16 as a function of the carrier frequency and coupling coefficient. The example system and its parameters are presented all together in Table 2.7. So far, in most of the presented examples, we assumed that both coils resonate at the carrier frequency. However, this is not always the case, such as in Fig. 2.16, where we have swept the carrier frequency from 12.5 to 14.5 MHz. The deductions of η_{Link} and P_{MN} as a function of the carrier frequency are presented in Table 2.8.

As can be seen in Fig. 2.16, when the link is strongly coupled, higher P_{MN} can be obtained operating at a frequency that is different from the resonant frequency of each coil.

Unlike P_{MN}, the frequency splitting effect does not occur in the PTE of the link, i.e., η_{Link} [24], which is always reduced if the carrier frequency is different from the resonance frequency of each coil even when the link is over-coupled, as shown in Fig. 2.17. This means that although the highest P_{MN} is obtained driving the system at one of the two natural frequencies, those frequencies are not the ones that achieve the highest η_{Link}.

Additionally, the frequency splitting effect in P_{MN} does not occur for all types of Tx-circuits. For instance, when a current source is used to drive a series resonant Tx coil as in Fig. 2.6, no frequency splitting effect occurs. This is because in that case, an increase in the reflected resistance increases the Tx-circuit output power, P_{Tx}, instead of reducing it.

Before trying to avoid the frequency splitting effect, we should verify if it is actually a problem in the WPT system. If it is possible to increase V_S, the reduction in P_{MN} can be simply addressed by increasing this voltage, using a closed-loop control. However, this may not always be possible or desirable, for instance, when the voltages across power amplifier transistors are near their breakdown level. When V_S cannot be increased, the most common solution is changing the carrier frequency in a way that it would follow the optimal P_{MN} path [25]. Another alternative could

Table 2.7 Frequency splitting effect in an example system. The calculation script is available in the supplementary material, file Sec217

Example circuit to visualize frequency splitting effect.

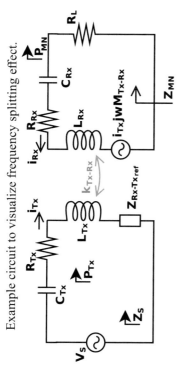

Fig. 2.15 WPT with series-series topology, used to address frequency splitting effect

In this example system, we used:

$$f_{res} = \frac{1}{2\pi\sqrt{L_{Tx}C_{Tx}}} = \frac{1}{2\pi\sqrt{L_{Rx}C_{Rx}}} = 13.56\,\text{MHz},\ Q_{Tx} = Q_{Rx} = 200,\ Q_L = 50,\ R_{Tx} = 5\,\Omega.\ \text{It should be mentioned that the coils quality factors}$$

also depend on the operating frequency, which is not considered in this analysis to simplify the discussion.

Case	Case I $V_S = 1\,\text{V}$ carrier, ω, at ω_{res} $k_{Tx\text{-}Rx} = 0.05$	Case II $V_S = 1\,\text{V}$ carrier, ω, at maximum P_{MN} $k_{Tx\text{-}Rx} = 0.05$	Case III $V_S = 1.74\,\text{V}$ carrier, ω, at ω_{res} $k_{Tx\text{-}Rx} = 0.05$
Z_S	$105\,\Omega$	$(25 - j10)$ ohms	$105\,\Omega$
P_{MN}	$3.6\,\text{mW}$	$11\,\text{mW}$	$11\,\text{mW}$
η_{Link}	76%	64%	76%

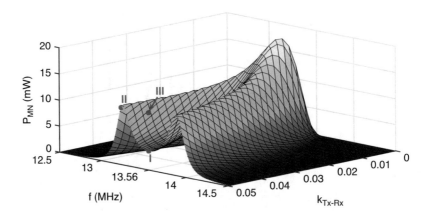

Fig. 2.16 Power delivered to the Rx-circuit, P_{MN}, as a function of $k_{Tx\text{-}Rx}$ and carrier frequency f, which deduction is presented in Table 2.8. $V_S = 1$ V and all the other parameters are summarized in Table 2.7. The calculation script is available in the supplementary material, file Sec217

be to size the coils in a way to avoid over-coupling even at the lowest link distance, $D_{Tx\text{-}Rx}$, as presented in [26, 27].

In Table 2.7 we compared three different Tx voltage sources in an over-coupled link. In case I, $V_S = 1$ V, and the carrier frequency is equal to the resonance frequency of each coil (Tx and Rx). It can be seen that P_{MN} can be increased by adjusting the carrier frequency to track one of the natural frequencies of the system, as presented in case II. However, this modification in the carrier frequency deteriorates the η_{Link}, as depicted in Fig. 2.17. In case III, instead of adjusting the carrier frequency, the voltage source is increased to have the same P_{MN} as in case II. By doing that, η_{Link} is not affected and remains at the same level that was obtained in case I. It is evident that increasing the voltage source, V_S, if possible, is a better solution than changing the carrier frequency because the link efficiency would not be compromised.

2.1.7.1 Analysis of Frequency Splitting Effect Based on T-Type Transformer Model

Interactions between two coupled coils can be modeled using their T-type equivalent circuit as shown in Fig. 2.18.

In [28] the frequency splitting effect is analyzed based on the equivalent T-type model. If the T-type equivalent model is substituted in Fig. 2.15, we obtain the circuit presented in Fig. 2.19. The results obtained using the T-type equivalent model (Fig. 2.19) are the same deduced from Fig. 2.15. In this subsection, we deduce the impedance seen by the Tx voltage source, Z_S, in (2.27) and compare it with the Z_S deduced in Table 2.8.

Table 2.8 Deduction of η_{Link} and P_{MN} as a function of the carrier frequency for the circuit presented in Fig. 2.15. It is used to generate Figs. 2.16 and 2.17 in the script available in the supplementary material, file Sec217

$$P_{MN} = \overbrace{\eta_{L_{Tx}}\eta_{L_{Rx}}}^{\eta_{Link}} \times P_{Tx}$$

$$P_{Tx} = \frac{1}{2}\left(\frac{V_S}{|Z_S|}\right)^2 Re\{Z_S\}$$

where,

$$\eta_{L_{Rx}} = \frac{Q_{Rx}}{Q_{Rx} + Q_L} \qquad (2.5)$$

$$\left.\begin{array}{l} Z_S = R_{Tx} + Z_{Rx\text{-}Tx_{ref}} + \\[4pt] \qquad + \dfrac{1}{j\omega C_{Tx}} + j\omega L_{Tx} \\[10pt] \dfrac{1}{j\omega C_{Tx}} = \dfrac{1}{j\omega C_{Tx}}\dfrac{j\omega L_{Tx}}{j\omega L_{Tx}} \\[10pt] = -\dfrac{1}{\omega^2}\underbrace{\dfrac{1}{C_{Tx}L_{Tx}}}_{\omega_{res}^2} j\omega L_{Tx} \end{array}\right\} \Rightarrow$$

$$\eta_{L_{Tx}} = \frac{Re\{Z_{Rx\text{-}Tx_{ref}}\}}{Re\{Z_{Rx\text{-}Tx_{ref}}\} + R_{Tx}}$$
$$(2.7)$$

$$\Rightarrow Z_S = R_{Tx} + Z_{Rx\text{-}Tx_{ref}} +$$

$$+ jR_{Tx}Q_{Tx}\left(1 - \frac{\omega_{res}^2}{\omega^2}\right)$$

$$(2.6) \Rightarrow Z_{Rx\text{-}Tx_{ref}} = \frac{\omega^2 M_{Tx\text{-}Rx}^2}{R_{Rx} + j\omega L_{Rx} + Z_{MN}}$$

$$= \frac{\omega^2 k_{Tx\text{-}Rx}^2 L_{Tx} L_{Rx}}{R_{Rx} + j\omega L_{Rx} + 1/(j\omega C_{Rx}) + R_L}$$

$$= \frac{\omega^2 k_{Tx\text{-}Rx}^2 L_{Tx} L_{Rx}}{R_{Rx} + R_L + j\omega L_{Rx}(1 - (\omega_{res}/\omega)^2)}$$

$$= \frac{k_{Tx\text{-}Rx}^2 \omega L_{Tx}}{\frac{R_{Rx}}{\omega L_{Rx}} + \frac{R_L}{\omega L_{Rx}} + j(1 - (\omega_{res}/\omega)^2)}$$

$$= \frac{k_{Tx\text{-}Rx}^2 \omega L_{Tx} Q_{Rx} Q_L}{Q_L + Q_{Rx} + jQ_{Rx}Q_L(1 - (\omega_{res}/\omega)^2)}$$

$$= \frac{k_{Tx\text{-}Rx}^2 Q_{Tx} Q_{Rx} Q_L}{Q_L + Q_{Rx} + jQ_{Rx}Q_L(1 - (\omega_{res}/\omega)^2)} R_{Tx}$$

$$(2.26)$$

$\omega_{res} = 1/\sqrt{C_{Tx}L_{Tx}} = 1/\sqrt{C_{Rx}L_{Rx}}$ is the resonance frequency of the Tx and Rx coils

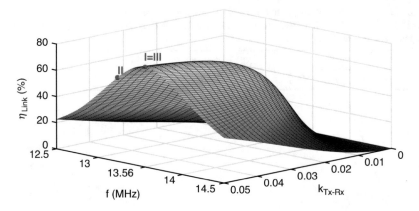

Fig. 2.17 PTE of the link, η_{Link}, as a function of $k_{Tx\text{-}Rx}$ and carrier frequency f, which deduction is presented in Table 2.8. $V_S = 1$ V and all the other parameters are summarized in Table 2.7. The calculation script is available in the supplementary material, file Sec217

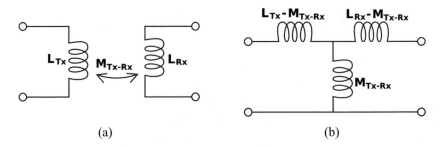

Fig. 2.18 Transformation to T-type equivalent model. (**a**) Typical two-port model of two coupled coils. (**b**) Equivalent T-type model

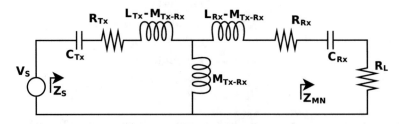

Fig. 2.19 WPT with series-series topology, used to address frequency splitting effect presented in Fig. 2.15, substituting the two coupled coils with its equivalent T-type model

$$Z_S = \frac{1}{j\omega C_{Tx}} + R_{Tx} + j\omega L_{Tx} - j\omega M_{Tx\text{-}Rx} +$$

$$+ j\omega M_{Tx\text{-}Rx} // \left(j\omega L_{Rx} - j\omega M_{Tx\text{-}Rx} + R_{Rx} + \frac{1}{j\omega C_{Rx}} + R_L \right) \Rightarrow$$

$$Z_S = \frac{1}{j\omega C_{Tx}} + R_{Tx} + j\omega L_{Tx} - j\omega M_{Tx\text{-}Rx} +$$

$$+ \frac{j\omega M_{Tx\text{-}Rx} \left(j\omega L_{Rx} - j\omega M_{Tx\text{-}Rx} + R_{Rx} + \frac{1}{j\omega C_{Rx}} + R_L \right)}{j\omega L_{Rx} + R_{Rx} + \frac{1}{j\omega C_{Rx}} + R_L} \Rightarrow$$

$$Z_S = \frac{1}{j\omega C_{Tx}} + R_{Tx} + j\omega L_{Tx} + \underbrace{\frac{(\omega M_{Tx\text{-}Rx})^2}{\underbrace{j\omega L_{Rx} + R_{Rx} + \frac{1}{j\omega C_{Rx}} + R_L}_{Z_{MN}}}}_{Z_{Rx\text{-}Tx_{ref}}} \Rightarrow$$

As can be seen, the Z_S deduced in (2.27) is the same as the one calculated in (2.27) Table 2.8.

If $M_{Tx\text{-}Rx}$ is much lower than L_{Tx} and L_{Rx} ($M_{Tx\text{-}Rx}$ tends towards zero), it is easy to see from Fig. 2.19 that Z_S is purely real when the carrier frequency is $\omega_{res} = 1/\sqrt{C_{Tx}L_{Tx}} = 1/\sqrt{C_{Rx}L_{Rx}}$. When $M_{Tx\text{-}Rx}$ is comparable to L_{Tx} and L_{Rx}, the system has two different natural frequencies, while Z_S continues being purely real at $\omega = \omega_{res} = 1/\sqrt{C_{Tx}L_{Tx}} = 1/\sqrt{C_{Rx}L_{Rx}}$. On the other hand, as was mentioned, if the carrier frequency is adapted to track one of the two natural frequencies of the system, $\omega \neq \omega_{res} = 1/\sqrt{C_{Tx}L_{Tx}} = 1/\sqrt{C_{Rx}L_{Rx}}$, the Z_S obtained is complex, e.g., case II (see Table 2.7). This result may not seem intuitive at first glance as the system's natural frequencies, which maximize the oscillating amplitudes and thus the P_{MN}, are different from the frequency that cancels the imaginary part of the system input impedance, Z_S.

It should be mentioned that in measurements, if the two coils are too close, the self-inductance may also be affected, the same way that it is affected by any foreign object, particularly if it is a conductive or ferromagnetic material. This shift in the self-inductance, which was not considered in this analysis, affects the frequency response of an over-coupled system even further than what the simple model that was presented here is predicting.

2.2 Reflected Load Theory in Systems with Additional Resonant Coils

2.2.1 Link Efficiency, η_{Link}, and Power Delivered to the Rx-circuit, P_{MN}, in a 3-Coil Link

As mentioned in Chap. 1, additional resonant coils can be used in the WPT link to increase η_{Link} and P_{MN} at a given Tx-Rx distance or extend the power transmission range for a desired minimum η_{Link} in a worst-case condition. In this section, we deduce the η_{Link}, and the received power P_{MN}, for a 3-coil link.

The simplified lumped circuit model for a 3-coil link is depicted in Fig. 2.20, where L_A and R_A are the additional resonator self-inductance and its parasitic resistance, respectively. The capacitor C_A was selected to ensure that the additional coil resonates at the carrier frequency. For the sake of simplicity, the coupling between non-neighboring inductors, i.e., $k_{Tx\text{-}Rx}$, has been considered negligible.

Analogous to Sect. 2.1.3, the efficiency of each coil is calculated by deducing the reflected impedance onto each coil from the one following it toward the load (Fig. 2.21).

The efficiency of the Rx coil is the same deduced in Sect. 2.1.3, as it does not depend on the previous coils, and it is:

$$\eta_{L_{Rx}} = \frac{Q_{Rx}}{Q_{Rx} + Q_L} = \frac{Q_{Rx\text{-}L}}{Q_L}, \tag{2.5}$$

where $Q_{Rx\text{-}L}$ (2.11) was used.

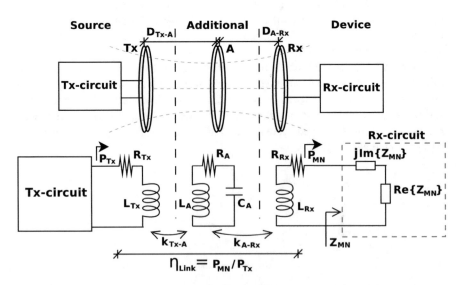

Fig. 2.20 A simplified diagram of a 3-coil WPT link and its lumped circuit model

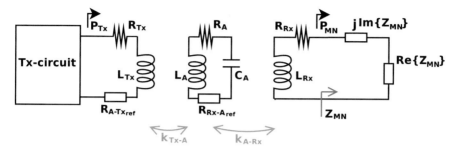

Fig. 2.21 3-coil link model including reflected resistance

Similar to (2.8), assuming a resonant Rx, $\omega L_{Rx} = -Im\{Z_{MN}\}$, the reflected resistance onto the additional resonator is:

$$R_{Rx\text{-}A_{ref}} = \frac{\omega^2 k_{A\text{-}Rx}{}^2 L_A L_{Rx}}{R_{Rx} + Re\{Z_{MN}\}} = \frac{k_{A\text{-}Rx}{}^2 Q_A Q_{Rx} Q_L}{Q_L + Q_{Rx}} R_A$$

$$= k_{A\text{-}Rx}{}^2 Q_A Q_{Rx\text{-}L} R_A, \tag{2.28}$$

where $k_{A\text{-}Rx}$ is the coupling coefficient between the additional resonator and the Rx coil and $Q_A = \frac{\omega L_A}{R_A}$. Therefore:

$$\eta_{LA} = \frac{R_{Rx\text{-}A_{ref}}}{R_{Rx\text{-}A_{ref}} + R_A} = \frac{k_{A\text{-}Rx}{}^2 Q_A Q_{Rx\text{-}L}}{k_{A\text{-}Rx}{}^2 Q_A Q_{Rx\text{-}L} + 1}. \tag{2.29}$$

The reflected resistance from the additional resonator onto the Tx coil ($R_{A\text{-}Tx_{ref}}$) is:

$$R_{A\text{-}Tx_{ref}} = \frac{\omega^2 k_{Tx\text{-}A}{}^2 L_{Tx} L_A}{R_{Rx\text{-}A_{ref}} + R_A}$$

$$= k_{Tx\text{-}A}{}^2 Q_{Tx} Q_A R_{Tx} \frac{R_A}{k_{A\text{-}Rx}{}^2 Q_A Q_{Rx\text{-}L} R_A + R_A} \tag{2.30}$$

$$= k_{Tx\text{-}A}{}^2 Q_{Tx} Q_A R_{Tx} \frac{1}{k_{A\text{-}Rx}{}^2 Q_A Q_{Rx\text{-}L} + 1}.$$

Finally the Tx efficiency is:

$$\eta_{LTx} = \frac{R_{A\text{-}Tx_{ref}}}{R_{A\text{-}Tx_{ref}} + R_{Tx}} = \frac{k_{Tx\text{-}A}{}^2 Q_{Tx} Q_A}{k_{Tx\text{-}A}{}^2 Q_{Tx} Q_A + k_{A\text{-}Rx}{}^2 Q_A Q_{Rx\text{-}L} + 1}. \tag{2.31}$$

And the total link efficiency ($\eta_{Link} = \eta_{LRx} \eta_{LA} \eta_{LTx}$) is:

$$\eta_{Link} = \frac{Q_{Rx\text{-}L}}{Q_L} \frac{k_{A\text{-}Rx}^2 Q_A Q_{Rx\text{-}L}}{(k_{A\text{-}Rx}^2 Q_A Q_{Rx\text{-}L} + 1)} \frac{k_{Tx\text{-}A}^2 Q_{Tx} Q_A}{(k_{Tx\text{-}A}^2 Q_{Tx} Q_A + k_{A\text{-}Rx}^2 Q_A Q_{Rx\text{-}L} + 1)}. \tag{2.32}$$

η_{Link} tends towards zero ($\eta_{Link} \rightarrow 0$) either when $k_{A\text{-}Rx}^2 Q_A Q_{Rx\text{-}L} \rightarrow 0$ or $k_{A\text{-}Rx}^2 Q_A Q_{Rx\text{-}L} \rightarrow \infty$. Therefore, an optimum value exists for $k_{A\text{-}Rx}^2 Q_A Q_{Rx\text{-}L}$:

$$(k_{A\text{-}Rx}^2 Q_A Q_{Rx\text{-}L})_{opt} = \sqrt{1 + k_{Tx\text{-}A}^2 Q_{Tx} Q_A}. \tag{2.33}$$

The condition in (2.33) can be fulfilled by adjusting the additional resonator's position or its coil geometry.

Finally, to calculate the output power, assuming the typical voltage source with a series resonant Tx:

$$\left.\begin{array}{c} P_{Tx} = \dfrac{V_S^2/2}{R_{Tx}+R_{A\text{-}Tx_{ref}}} \\ (2.30) \end{array}\right\} \Rightarrow \left.\begin{array}{c} P_{Tx} = \dfrac{V_S^2}{2R_{Tx}} \dfrac{k_{A\text{-}Rx}^2 Q_A Q_{Rx\text{-}L}+1}{k_{Tx\text{-}A}^2 Q_{Tx} Q_A + k_{A\text{-}Rx}^2 Q_A Q_{Rx\text{-}L}+1} \\ (2.32) \end{array}\right\} \Rightarrow$$

$$\Rightarrow P_{MN} = \eta_{Link} P_{Tx} = \frac{V_S^2}{2R_{Tx}} \frac{Q_{Rx\text{-}L}}{Q_L} \frac{(k_{A\text{-}Rx}^2 Q_A Q_{Rx\text{-}L})(k_{Tx\text{-}A}^2 Q_{Tx} Q_A)}{(k_{Tx\text{-}A}^2 Q_{Tx} Q_A + k_{A\text{-}Rx}^2 Q_A Q_{Rx\text{-}L} + 1)^2}. \tag{2.34}$$

Analogous to η_{Link}, P_{MN} also tends towards zero ($P_{MN} \rightarrow 0$) either when $k_{A\text{-}Rx}^2 Q_A Q_{Rx\text{-}L} \rightarrow 0$ or $k_{A\text{-}Rx}^2 Q_A Q_{Rx\text{-}L} \rightarrow \infty$. Therefore, there also exists an optimum value for $k_{A\text{-}Rx}^2 Q_A Q_{Rx\text{-}L}$ that maximizes (2.34), which is presented in (2.35):

$$(k_{A\text{-}Rx}^2 Q_A Q_{Rx\text{-}L})_{opt} = 1 + k_{Tx\text{-}A}^2 Q_{Tx} Q_A. \tag{2.35}$$

If the driver is a current source, I_S, with a series resonant Tx, (2.34) should be modified as follows:

$$\left.\begin{array}{c} P_{Tx} = \dfrac{I_S^2}{2}(R_{Tx} + R_{Rx\text{-}Tx_{ref}}) \\ (2.30) \end{array}\right\} \Rightarrow$$

$$\left.\begin{array}{c} \Rightarrow P_{Tx} = \dfrac{I_S^2 R_{Tx}}{2} \dfrac{(k_{Tx\text{-}A}^2 Q_{Tx} Q_A+k_{A\text{-}Rx}^2 Q_A Q_{Rx\text{-}L}+1)}{k_{A\text{-}Rx}^2 Q_A Q_{Rx\text{-}L}+1} \\ (2.32) \end{array}\right\} \Rightarrow$$

$$\Rightarrow P_{MN} = \eta_{Link} P_{Tx} = \frac{I_S^2 R_{Tx}}{2} \frac{Q_{Rx\text{-}L}}{Q_L} \frac{(k_{A\text{-}Rx}^2 Q_A Q_{Rx\text{-}L})(k_{Tx\text{-}A}^2 Q_{Tx} Q_A)}{(k_{A\text{-}Rx}^2 Q_A Q_{Rx\text{-}L} + 1)^2}. \tag{2.36}$$

2.2.2 Generalization to N-Coil Links

In this section, η_{Link} and P_{MN} are calculated for the N-coil resonant link depicted in Fig. 2.22. For the sake of simplicity, the coupling between non-neighboring inductors has been considered negligible.

As was done for the 2-coil link in Sect. 2.1.3 and for the 3-coil link in Sect. 2.2.1, the η_{Link} can be calculated as the product of the efficiency of each coil. The efficiency of coil i is calculated as the ratio between the power transferred to the next coil, $i + 1$, and the power received by the coil i. To do that, it is necessary to calculate the reflected resistance in each coil. Analogous to (2.6), the reflected resistance in the coil i from the coil $i + 1$ is deduced in (2.37). In this section, we refer to the reflected resistance in coil i from coil $i + 1$ as R_{Li}, because it is the resistance that loads the coil i:

$$
\left.\begin{array}{l}
V_{L_i} = j\omega L_i i_i + j\omega M_{i\text{-}(i+1)} i_{(i+1)} \\[4pt]
i_{(i+1)} = \dfrac{-i_i\, j\omega M_{i\text{-}(i+1)}}{R_{(i+1)} + j\omega L_{(i+1)} + \frac{1}{j\omega C_{(i+1)}} + R_{L(i+1)}}
\end{array}\right\} \Longrightarrow
$$

$$
\Longrightarrow \frac{V_{L_i}}{i_i} = j\omega L_i + \underbrace{\frac{\omega^2 M_{i\text{-}(i+1)}{}^2}{R_{(i+1)} + \cancel{j\omega L_{(i+1)}} + \cancel{\frac{1}{j\omega C_{(i+1)}}} + R_{L(i+1)}}}_{R_{Li}} \Longrightarrow
$$

$$
\Longrightarrow R_{Li} = \frac{\omega^2 k_{i\text{-}(i+1)}{}^2 L_i L_{(i+1)}}{R_{(i+1)} + R_{L(i+1)}} = k_{i\text{-}(i+1)}{}^2 \underbrace{\frac{\omega L_i}{R_i}}_{Q_i} R_i \underbrace{\frac{\omega L_{(i+1)}}{R_{(i+1)} + R_{L(i+1)}}}_{Q_{(i+1)L}} \Longrightarrow
$$

$$
\Longrightarrow R_{Li} = k_{i\text{-}(i+1)}{}^2 Q_i Q_{(i+1)L} R_i
\tag{2.37}
$$

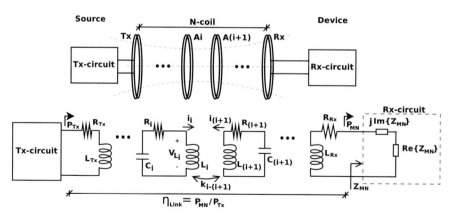

Fig. 2.22 A simplified diagram of an N-coil WPT link and its lumped circuit model

Therefore, in an N-coil link, each coil is loaded with an R_{Li}:

$$R_{Li} = k_{i\text{-}(i+1)}{}^2 Q_i Q_{(i+1)L} R_i, \text{ when } i = 1, 2, \ldots, (n-1)$$
$$R_{Ln} = Re\{Z_{MN}\}$$

$$\text{(2.38)}$$

where:

$$Q_i = \frac{\omega L_i}{R_i}$$

$$Q_{iL} = \frac{\omega L_i}{R_i + R_{Li}}.$$

$$\text{(2.39)}$$

The efficiency of each coil is:

$$\eta_i = \frac{R_{Li}}{R_i + R_{Li}}$$

$$\text{(2.40)}$$

and the efficiency of the N-coil link is:

$$\eta_{Link} = \prod_{i=1}^{n} \eta_i.$$

$$\text{(2.41)}$$

Assuming a voltage source Tx with a series resonant capacitor, the P_{MN} can be calculated as:

$$P_{MN} = \frac{V_S{}^2}{2(R_1 + R_{L1})} \eta_{Link}$$

$$\text{(2.42)}$$

2.2.3 Link Efficiency, η_{Link}, and Power Delivered to the Rx-circuit, P_{MN}, in a 4-Coil Link

The link efficiency, η_{Link}, and power delivered to the Rx-circuit, P_{MN}, in a 4-coil link (Fig. 2.23) are calculated in Table 2.9 applying the generalization for the N-coil link deduced in Sect. 2.2.2.

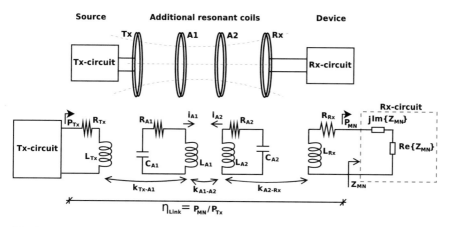

Fig. 2.23 A simplified diagram of a 4-coil WPT link and its lumped circuit model

Table 2.9 Deduction of η_{Link} and P_{MN} for a 4-coil link using the generalization deduced in Sect. 2.2.2

i	R_{Li} using (2.38)	$Q_{i\text{-}L}$ using (2.39)	η_i using (2.40)
4	$R_{L4} = Re\{Z_{MN}\}$	$Q_{Rx\text{-}L} = \dfrac{\omega L_{Rx}}{R_{Rx}+Re\{Z_{MN}\}}$	$\eta_{Rx} = \dfrac{Re\{Z_{MN}\}}{R_{Rx}+Re\{Z_{MN}\}} = \dfrac{Q_{Rx\text{-}L}}{Q_L}$
3	$R_{L3} = k_{A2\text{-}Rx}{}^2 Q_{A2} Q_{Rx\text{-}L} R_{A2}$	$Q_{A2\text{-}L} = \underbrace{\dfrac{\omega L_{A2}}{R_{A2}}}_{Q_{A2}} \dfrac{1}{1+k_{A2\text{-}Rx}{}^2 Q_{A2} Q_{Rx\text{-}L}}$	$\eta_{A2} = \dfrac{k_{A2\text{-}Rx}{}^2 Q_{A2} Q_{Rx\text{-}L}}{1+k_{A2\text{-}Rx}{}^2 Q_{A2} Q_{Rx\text{-}L}}$
2	$R_{L2} = k_{A1\text{-}A2}{}^2 Q_{A1} Q_{A2\text{-}L} R_{A1}$	$Q_{A1\text{-}L} = \underbrace{\dfrac{\omega L_{A1}}{R_{A1}}}_{Q_{A1}} \dfrac{1}{1+k_{A1\text{-}A2}{}^2 Q_{A1} Q_{A2\text{-}L}}$	$\eta_{A1} = \dfrac{k_{A1\text{-}A2}{}^2 Q_{A1} Q_{A2\text{-}L}}{1+k_{A1\text{-}A2}{}^2 Q_{A1} Q_{A2\text{-}L}}$
1	$R_{L1} = k_{Tx\text{-}A1}{}^2 Q_{Tx} Q_{A1\text{-}L} R_{Tx}$	$Q_{Tx\text{-}L} = \underbrace{\dfrac{\omega L_{Tx}}{R_{Tx}}}_{Q_{Tx}} \dfrac{1}{1+k_{Tx\text{-}A1}{}^2 Q_{Tx} Q_{A1\text{-}L}}$	$\eta_{Tx} = \dfrac{k_{Tx\text{-}A1}{}^2 Q_{Tx} Q_{A1\text{-}L}}{1+k_{Tx\text{-}A1}{}^2 Q_{Tx} Q_{A1\text{-}L}}$

Table 2.9 (continued)

$$\eta_{Link} = \eta_{Rx}\eta_{A2}\eta_{A1}\eta_{Tx} = \frac{Q_{Rx\text{-}L}}{Q_L}\frac{k_{A2\text{-}Rx}^2 Q_{A2}Q_{Rx\text{-}L}}{(1+k_{A2\text{-}Rx}^2 Q_{A2}Q_{Rx\text{-}L})}\frac{k_{A1\text{-}A2}^2 Q_{A1}Q_{A2\text{-}L}}{(1+k_{A1\text{-}A2}^2 Q_{A1}Q_{A2\text{-}L})}\frac{k_{Tx\text{-}A1}^2 Q_{Tx}Q_{A1\text{-}L}}{(1+k_{Tx\text{-}A1}^2 Q_{Tx}Q_{A1\text{-}L})} \Rightarrow$$

$$\eta_{Link} = \frac{Q_{Rx\text{-}L}}{Q_L}\frac{k_{A2\text{-}Rx}^2 Q_{A2}Q_{Rx\text{-}L}}{(1+k_{A2\text{-}Rx}^2 Q_{A2}Q_{Rx\text{-}L})}\frac{k_{A1\text{-}A2}^2 Q_{A1}\frac{Q_{A2}}{1+k_{A2\text{-}Rx}^2 Q_{A2}Q_{Rx\text{-}L}}}{(1+k_{A1\text{-}A2}^2 Q_{A1}Q_{A2\text{-}L})}\frac{k_{Tx\text{-}A1}^2 Q_{Tx}\frac{Q_{A1}}{1+k_{A1\text{-}A2}^2 Q_{A1}Q_{A2\text{-}L}}}{(1+k_{Tx\text{-}A1}^2 Q_{Tx}Q_{A1\text{-}L})} \Rightarrow$$

$$\eta_{Link} = \frac{Q_{Rx\text{-}L}}{Q_L}\frac{k_{A2\text{-}Rx}^2 Q_{A2}Q_{Rx\text{-}L}}{(1+k_{A2\text{-}Rx}^2 Q_{A2}Q_{Rx\text{-}L})^2}\frac{k_{A1\text{-}A2}^2 Q_{A1}Q_{A2}}{(1+k_{A1\text{-}A2}^2 Q_{A1}Q_{A2\text{-}L})^2}\frac{k_{Tx\text{-}A1}^2 Q_{Tx}Q_{A1}}{(1+k_{Tx\text{-}A1}^2 Q_{Tx}Q_{A1\text{-}L})} \Rightarrow$$

$$\eta_{Link} = \frac{Q_{Rx\text{-}L}}{Q_L}\frac{k_{A2\text{-}Rx}^2 Q_{A2}Q_{Rx\text{-}L}}{(1+k_{A2\text{-}Rx}^2 Q_{A2}Q_{Rx\text{-}L})^2}\frac{k_{A1\text{-}A2}^2 Q_{A1}Q_{A2}}{(1+k_{A1\text{-}A2}^2 Q_{A1}Q_{A2\text{-}L})^2}\frac{k_{Tx\text{-}A1}^2 Q_{Tx}Q_{A1}}{\left(1+k_{Tx\text{-}A1}^2 Q_{Tx}\frac{Q_{A1}}{1+k_{A1\text{-}A2}^2 Q_{A1}Q_{A2\text{-}L}}\right)} \Rightarrow$$

$$\eta_{Link} = \frac{Q_{Rx\text{-}L}}{Q_L}\frac{k_{A2\text{-}Rx}^2 Q_{A2}Q_{Rx\text{-}L}}{(1+k_{A2\text{-}Rx}^2 Q_{A2}Q_{Rx\text{-}L})^2}\frac{k_{A1\text{-}A2}^2 Q_{A1}Q_{A2}}{(1+k_{A1\text{-}A2}^2 Q_{A1}Q_{A2\text{-}L})}\frac{k_{Tx\text{-}A1}^2 Q_{Tx}Q_{A1}}{(1+k_{A1\text{-}A2}^2 Q_{A1}Q_{A2\text{-}L}+k_{Tx\text{-}A1}^2 Q_{Tx}Q_{A1})} \Rightarrow$$

$$\eta_{Link} = \frac{Q_{Rx\text{-}L}}{Q_L}\frac{k_{A2\text{-}Rx}^2 Q_{A2}Q_{Rx\text{-}L}}{(1+k_{A2\text{-}Rx}^2 Q_{A2}Q_{Rx\text{-}L})^2}\cdot$$

$$\cdot\frac{k_{A1\text{-}A2}^2 Q_{A1}Q_{A2}}{\left(1+k_{A1\text{-}A2}^2 Q_{A1}\frac{Q_{A2}}{1+k_{A2\text{-}Rx}^2 Q_{A2}Q_{Rx\text{-}L}}\right)}\frac{k_{Tx\text{-}A1}^2 Q_{Tx}Q_{A1}}{\left(1+k_{A1\text{-}A2}^2 Q_{A1}\frac{Q_{A2}}{1+k_{A2\text{-}Rx}^2 Q_{A2}Q_{Rx\text{-}L}}+k_{Tx\text{-}A1}^2 Q_{Tx}Q_{A1}\right)} \Rightarrow$$

$$\eta_{Link} = \frac{Q_{Rx\text{-}L}}{Q_L}\frac{k_{A2\text{-}Rx}^2 Q_{A2}Q_{Rx\text{-}L}k_{A1\text{-}A2}^2 Q_{A1}Q_{A2}k_{Tx\text{-}A1}^2 Q_{Tx}Q_{A1}}{[1+k_{A2\text{-}Rx}^2 Q_{A2}Q_{Rx\text{-}L}+k_{A1\text{-}A2}^2 Q_{A1}Q_{A2}][k_{A1\text{-}A2}^2 Q_{A1}Q_{A2}+(1+k_{Tx\text{-}A1}^2 Q_{Tx}Q_{A1})(1+k_{A2\text{-}Rx}^2 Q_{A2}Q_{Rx\text{-}L})]} \tag{2.43}$$

Voltage source and series resonant Tx $\Rightarrow P_{MN} = \dfrac{V_S^2}{2(R_1+R_{L1})}\eta_{Link} = \dfrac{V_S^2}{2R_{Tx}(1+k_{Tx\text{-}A1}^2 Q_{Tx}Q_{A1\text{-}L})}\eta_{Link} \Rightarrow$

$$P_{MN} = \frac{V_S^2}{2R_{Tx}}\frac{1}{\left(1+k_{Tx\text{-}A1}^2 Q_{Tx}\frac{Q_{A1}}{1+k_{A1\text{-}A2}^2 Q_{A1}\frac{Q_{A2}}{1+k_{A2\text{-}Rx}^2 Q_{A2}Q_{Rx\text{-}L}}}\right)}\eta_{Link} \Rightarrow$$

$$P_{MN} = \frac{V_S^2}{2R_{Tx}}\frac{1}{\left(1+\frac{k_{Tx\text{-}A1}^2 Q_{Tx}Q_{A1}(1+k_{A2\text{-}Rx}^2 Q_{A2}Q_{Rx\text{-}L})}{1+k_{A2\text{-}Rx}^2 Q_{A2}Q_{Rx\text{-}L}+k_{A1\text{-}A2}^2 Q_{A1}Q_{A2}}\right)}\eta_{Link} \Rightarrow$$

$$P_{MN} = \frac{V_S^2}{2R_{Tx}}\frac{1+k_{A2\text{-}Rx}^2 Q_{A2}Q_{Rx\text{-}L}+k_{A1\text{-}A2}^2 Q_{A1}Q_{A2}}{[1+k_{A2\text{-}Rx}^2 Q_{A2}Q_{Rx\text{-}L}+k_{A1\text{-}A2}^2 Q_{A1}Q_{A2}+k_{Tx\text{-}A1}^2 Q_{Tx}Q_{A1}(1+k_{A2\text{-}Rx}^2 Q_{A2}Q_{Rx\text{-}L})]}\eta_{Link} \Rightarrow$$

$$P_{MN} = \frac{V_S^2}{2R_{Tx}}\frac{1+k_{A2\text{-}Rx}^2 Q_{A2}Q_{Rx\text{-}L}+k_{A1\text{-}A2}^2 Q_{A1}Q_{A2}}{[k_{A1\text{-}A2}^2 Q_{A1}Q_{A2}+(1+k_{Tx\text{-}A1}^2 Q_{Tx}Q_{A1})(1+k_{A2\text{-}Rx}^2 Q_{A2}Q_{Rx\text{-}L})]}\eta_{Link} \Rightarrow$$

$$P_{MN} = \frac{V_S^2}{2R_{Tx}}\frac{Q_{Rx\text{-}L}}{Q_L}\frac{k_{A2\text{-}Rx}^2 Q_{A2}Q_{Rx\text{-}L}k_{A1\text{-}A2}^2 Q_{A1}Q_{A2}k_{Tx\text{-}A1}^2 Q_{Tx}Q_{A1}}{[k_{A1\text{-}A2}^2 Q_{A1}Q_{A2}+(1+k_{Tx\text{-}A1}^2 Q_{Tx}Q_{A1})(1+k_{A2\text{-}Rx}^2 Q_{A2}Q_{Rx\text{-}L})]^2} \tag{2.44}$$

2.3 Comparison Between 2-, 3-, and 4-Coil Links

In order to provide intuition about where each topology is better and why, this section starts comparing the three example systems presented in Fig. 2.24. The parameters used are summarized in Table 2.10. The η_{Link} of the 2-coil link is

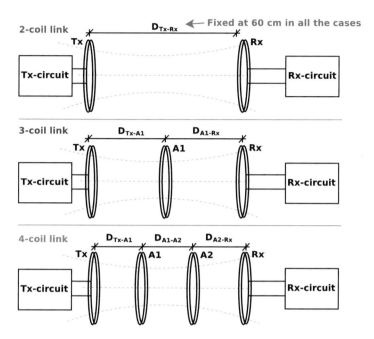

Fig. 2.24 The 2-, 3-, and 4-coil links compared. The $D_{Tx\text{-}Rx} = 60$ cm is kept constant for the three cases

Table 2.10 Parameters used to compare 2-, 3-, and 4-coil links (Figs. 2.25 and 2.26)

Coils	All identical, with: $Q_{Tx} = Q_{A1} = Q_{A2} = Q_{Rx} = 200$, $R_{Tx} = 5\,\Omega$, circular shape, diameter 20 cm
Distance	$D_{Tx\text{-}Rx} = 60$ cm, aligned
Coupling coefficient	is calculated using the approximation (3.16), Chap. 3
Tx-circuit	$V_S = 1$ V series resonant capacitor
Rx-circuit	$Q_L = 50$

calculated using (2.13) and is 11%. The η_{Link} of the 3-coil coil link is plotted in Fig. 2.25, using (2.32), as a function of the distance between the Tx and the additional resonant coil. The η_{Link} of the 4-coil link is also plotted in Fig. 2.25, using (2.43), as a function of the distance between the Tx and the first resonator, A1, for two different distances (10 and 20 cm) between Rx and the second resonator, A2. Under the same conditions, P_{MN} was calculated and presented in Fig. 2.26.

As can be seen in Figs. 2.25 and 2.26, much higher η_{Link} and P_{MN} can be achieved by adding resonators. However, the improvements in both, η_{Link} and P_{MN}, depend on the position of the resonator, and for some positions, instead of improving the performance, it is degraded.

The resonator can be thought of as a passive relay. Although it relays the power carrier signal, it also introduces new losses due to its parasitic resistance; thus adding a relay may not always be beneficial. In the 3-coil example system presented in Fig. 2.25, the maximum η_{Link} is obtained when the additional resonator (relay) is approximately equidistant to Tx and Rx. That position which maximizes η_{Link} (2.32) corresponds to the position that fulfills (2.33), as deduced in Sect. 2.2.1. As shown in Fig. 2.26, the P_{MN} of the 3-coil link, for this case, deduced in (2.34), is maximized for a slightly different additional resonator position, which satisfies (2.35). As was proved in (2.33) and (2.35), the positions that maximize η_{Link} and P_{MN} depend on $Q_{Rx\text{-}L}$ which depends on the Rx-circuit that loads the Rx coil.

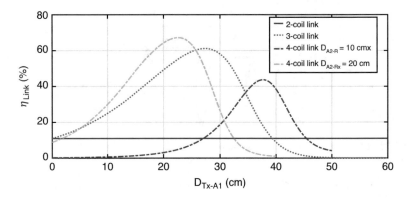

Fig. 2.25 η_{Link} for a 2-coil link using (2.13), for a 3-coil link using (2.32), and for a 4-coil link using (2.43). The parameters used are presented in Table 2.10. The calculation script is available in the supplementary material, file Sec230

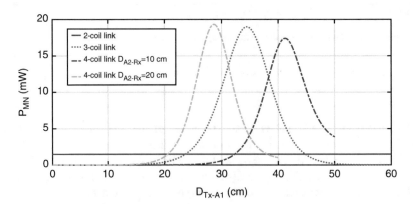

Fig. 2.26 P_{MN} for a 2-coil link using (2.17), for a 3-coil link using (2.34), and for a 4-coil link using (2.44). The parameters used are presented in Table 2.10. The calculation script is available in the supplementary material, file Sec230

Therefore, if the Rx-circuit power consumption changes, the theoretical optimal position of the additional coil could be altered.

Regarding the 4-coil link, as shown in Fig. 2.25, an η_{Link} higher than the 2- and the 3-coil links can be achieved. However, the resonators should be correctly placed; otherwise, the efficiency will not be improved. This optimal positions of the additional resonators also depend on the Rx-circuit input impedance, Z_{MN}.

As discussed in Sects. 2.1.5 and 2.1.7, when there is a resonant coil near the Tx, the reflected impedance onto the Tx coil increases, which reduces the Tx-circuit output power, P_{Tx}. This is the reason why P_{MN} drops to almost zero when the first additional resonator, A1, is placed too close to the Tx coil in Fig. 2.26.

Depending on the application, it may not be possible to place the resonator in between Tx and Rx. Additionally, the inclusion of a resonator may not be useful even if its position is optimized. Figure 2.27 was generated to exemplify this fact. In Fig. 2.27, a 2-coil link is compared against a 3-coil link in which additional resonator is placed in the optimal position maximizing η_{Link}. The optimal resonator position for this example 3-coil link is plotted in Fig. 2.28. The parameters used here are the same as presented in Table 2.10, but in Fig. 2.27, the Tx to Rx distance, $D_{Tx\text{-}Rx}$, is swept. As can be seen from Fig. 2.27, in this example, the inclusion of an identical resonant coil between Tx and Rx is not beneficial for distances $D_{Tx\text{-}Rx} < 20$ cm. However, for larger distances, the improvements are significant. For instance, at $D_{Tx\text{-}Rx} = 60$ cm, the efficiency is improved from $\eta_{Link} = 11\%$ to $\eta_{Link} \simeq 60\%$, thanks to the additional resonant coil.

The effect observed in Fig. 2.27 can be predicted by comparing the η_{Link} of 2 and 3-coil links. Comparing the 2-coil (2.13) and 3-coil (2.32) link efficiencies, it can be seen that the Rx coil efficiency ($Q_{Rx\text{-}L}/Q_L$) is present in both expressions, and therefore, it does not affect this comparison. Defining $\mathbb{K}_{Tx\text{-}Rx\text{-}L} = k_{Tx\text{-}Rx}{}^2 Q_{Tx} Q_{Rx\text{-}L}$ and $\mathbb{K}_{Tx\text{-}A} = k_{Tx\text{-}A}{}^2 Q_{Tx} Q_A$ and assuming $(k_{A\text{-}Rx}{}^2 Q_A Q_{Rx\text{-}L})_{opt}$ (2.33) is adopted, i.e., the optimal additional resonator position is selected for the additional resonator,

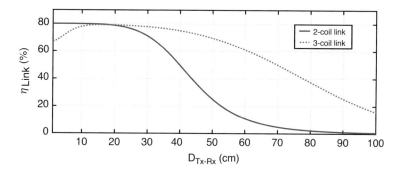

Fig. 2.27 Comparison between 2-coil and 3-coil links. The distance between Tx and Rx is identical for both systems, and the additional resonator position is selected to maximize η_{Link}. All the specs of the example system used to generate this plot are summarized in Table 2.10. The calculation script is available in the supplementary material, file Sec230

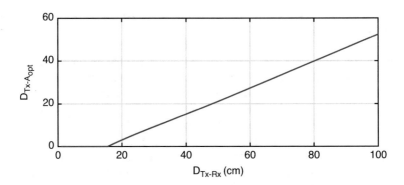

Fig. 2.28 The optimal additional resonator position that maximizes η_{3coil} as a function of $D_{Tx\text{-}Rx}$. This optimal position is the one used for each $D_{Tx\text{-}Rx}$ in Fig. 2.27. The calculation script is available in the supplementary material, file Sec230

the 2-coil and 3-coil link efficiencies can be expressed as:

$$\eta_{2coil} = \frac{Q_{Rx\text{-}L}}{Q_L} \frac{\mathbb{K}_{Tx\text{-}Rx\text{-}L}}{\mathbb{K}_{Tx\text{-}Rx\text{-}L} + 1} = \frac{Q_{Rx\text{-}L}}{Q_L} f_2(\mathbb{K}_{Tx\text{-}Rx\text{-}L})$$

$$\eta_{3coil} = \frac{Q_{Rx\text{-}L}}{Q_L} \frac{\sqrt{1 + \mathbb{K}_{Tx\text{-}A}} \cdot \mathbb{K}_{Tx\text{-}A}}{(1 + \sqrt{1 + \mathbb{K}_{Tx\text{-}A}})(1 + \mathbb{K}_{Tx\text{-}A} + \sqrt{1 + \mathbb{K}_{Tx\text{-}A}})} = \frac{Q_{Rx\text{-}L}}{Q_L} f_3(\mathbb{K}_{Tx\text{-}A})$$

$$(2.45)$$

From (2.45), if we assume that the $\mathbb{K}_{Tx\text{-}Rx\text{-}L}$ of the 2-coil link is equal to the $\mathbb{K}_{Tx\text{-}A}$ of the 3-coil link, $\mathbb{K}_{Tx\text{-}Rx\text{-}L} = \mathbb{K}_{Tx\text{-}A}$, thus $\eta_{2coil} > \eta_{3coil}$, which means that the inclusion of an additional resonator under this condition has no benefits and can reduce η_{Link}.

The larger $\mathbb{K}_{Tx\text{-}A}$ or $\mathbb{K}_{Tx\text{-}Rx\text{-}L}$ are, the higher the efficiencies, as f_2 and f_3 are monotonically increasing functions. Therefore, the efficiency can be improved by an additional resonator if $\mathbb{K}_{Tx\text{-}A}$ is sufficiently higher than $\mathbb{K}_{Tx\text{-}Rx\text{-}L}$ in order to obtain $f_3(\mathbb{K}_{Tx\text{-}A}) > f_2(\mathbb{K}_{Tx\text{-}Rx\text{-}L})$ (2.45). To achieve $\mathbb{K}_{Tx\text{-}A} > \mathbb{K}_{Tx\text{-}Rx\text{-}L}$, the additional resonator should be closer to the Tx or has a larger radius than the receiver ($k_{Tx\text{-}A} > k_{Tx\text{-}Rx}$). Additionally, $\mathbb{K}_{Tx\text{-}A} > \mathbb{K}_{Tx\text{-}Rx\text{-}L}$ can be achieved if $Q_A > Q_{Rx\text{-}L}$, which is possible because Q_A is determined by the parasitic resistance of the additional resonator, while $Q_{Rx\text{-}L}$ (2.11) has an external load resistor. This is indeed one of the most important advantages of 3-coil links over 2-coil links.

Appendices

A.1 P_{MN} Calculation for a Voltage Source and Series Tx Resonance

From Fig. 2.4, with a sinusoidal voltage source with a peak value of V_S, and a series resonant Tx $1/(C_{Tx}\omega) = \omega L_{Tx}$,

$$\left. P_{Tx} = \frac{V_S^2/2}{R_{Tx}+R_{Rx\text{-}Tx_{ref}}} \right\} \implies \left. P_{Tx} = \frac{V_S^2/2}{R_{Tx}(k_{Tx\text{-}Rx}^2 Q_{Tx}Q_{Rx\text{-}L}+1)} \right\} \implies$$

$$(2.10) \qquad\qquad\qquad\qquad (2.13) \qquad\qquad\qquad (2.17)$$

$$\implies P_{MN} = \eta_{Link}P_{Tx} = \frac{V_S^2}{2R_{Tx}} \frac{Q_{Rx\text{-}L}}{Q_L} \frac{k_{Tx\text{-}Rx}^2 Q_{Tx}Q_{Rx\text{-}L}}{(k_{Tx\text{-}Rx}^2 Q_{Tx}Q_{Rx\text{-}L}+1)^2}$$

A.2 P_{MN} Calculation for a Voltage Source and Parallel Tx Resonance

From Fig. 2.5, with a voltage source and a parallel resonant Tx, $1/(C_{Tx}\omega) = \omega L_{Tx}$, the source load impedance, Z_S, can be calculated as

$$\left. Z_S = \frac{1}{j\omega C_{Tx}+1/(j\omega L_{Tx}+R_{Tx}+R_{Rx\text{-}Tx_{ref}})} = \frac{j\omega L_{Tx}+R_{Tx}+R_{Rx\text{-}Tx_{ref}}}{j\omega C_{Tx}(j\omega L_{Tx}+R_{Tx}+R_{Rx\text{-}Tx_{ref}})+1} \atop \text{Resonance} \Rightarrow 1/(C_{Tx}\omega) = \omega L_{Tx} \right\} \implies$$

$$\left. \Rightarrow Z_S = \frac{j\omega L_{Tx}+R_{Tx}+R_{Rx\text{-}Tx_{ref}}}{j\omega C_{Tx}(R_{Tx}+R_{Rx\text{-}Tx_{ref}})} \atop ^1\text{assuming } \omega L_{Tx} \gg R_{Tx}+R_{Rx\text{-}Tx_{ref}} \right\} \implies$$

$$\left. \Rightarrow Z_S = \frac{L_{Tx}}{C_{Tx}(R_{Tx}+R_{Rx\text{-}Tx_{ref}})} \atop \text{Resonance} \Rightarrow 1/(C_{Tx}\omega) = \omega L_{Tx} \Rightarrow C_{Tx} = 1/(\omega^2 L_{Tx}) \right\} \implies$$

$$\Rightarrow Z_S = \frac{(\omega L_{Tx})^2}{R_{Tx}+R_{Rx\text{-}Tx_{ref}}}.$$
$$(2.46)$$

From (2.46), P_{MN} is

---

[1] Note that if $Q_{Tx} \gg 1 \Rightarrow \omega L_{Tx} \gg R_{Tx}$. Thus assuming that the reflected resistance, $R_{Rx\text{-}Tx_{ref}}$ (2.10), is not too large, the assumption is reasonable. Note that if it is not the case, the Tx resonance is affected by variations in the reflected impedance which is not desired.

$$P_{Tx} = \frac{1}{2}\frac{V_S^2}{|Z_S|^2}Re\{Z_S\} = \frac{V_S^2}{2|Z_S|} = \frac{V_S^2}{2(\omega L_{Tx})^2}(R_{Tx} + R_{Rx\text{-}Tx_{ref}}) \left.\begin{array}{c} \\ (2.10) \end{array}\right\} \Rightarrow$$

$$\Rightarrow P_{Tx} = \frac{V_S^2}{2(\omega L_{Tx})^2}R_{Tx}(k_{Tx\text{-}Rx}^2 Q_{Tx}Q_{Rx\text{-}L} + 1) \left.\begin{array}{c} \\ (2.13) \end{array}\right\} \Rightarrow \qquad (2.18)$$

$$\Rightarrow P_{MN} = \eta_{Link}P_{Tx} = \frac{V_S^2}{2(\omega L_{Tx})^2}R_{Tx}\frac{Q_{Rx\text{-}L}}{Q_L}k_{Tx\text{-}Rx}^2 Q_{Tx}Q_{Rx\text{-}L}$$

This case is presented with a parallel capacitor to preserve the analogy with the other cases. However, the parallel capacitor, like any other element in parallel with the source, is not affecting the results as the source impedance has been ignored.

A.3 P_{MN} Calculation for a Current Source and Series Tx Resonance

From Fig. 2.6, with a current source and a series resonant Tx

$$P_{Tx} = \frac{I_S^2}{2}(R_{Tx} + R_{Rx\text{-}Tx_{ref}}) \left.\begin{array}{c} \\ (2.10) \end{array}\right\} \Rightarrow$$

$$\Rightarrow P_{Tx} = \frac{I_S^2}{2}R_{Tx}(k_{Tx\text{-}Rx}^2 Q_{Tx}Q_{Rx\text{-}L} + 1) \left.\begin{array}{c} \\ (2.13) \end{array}\right\} \Rightarrow \qquad (2.19)$$

$$\Rightarrow P_{MN} = \eta_{Link}P_{Tx} = \frac{I_S^2}{2}R_{Tx}\frac{Q_{Rx\text{-}L}}{Q_L}k_{Tx\text{-}Rx}^2 Q_{Tx}Q_{Rx\text{-}L}$$

A.4 P_{MN} Calculation for a Current Source and Parallel Tx Resonance

From Fig. 2.7, and using the expression of Z_S already calculated in (2.46)

$$P_{Tx} = \frac{I_S^2}{2}Re\{Z_S\} = \frac{1}{2}\frac{I_S^2(\omega L_{Tx})^2}{R_{Tx}+R_{Rx\text{-}Tx_{ref}}} \left.\begin{array}{c} \\ (2.10) \end{array}\right\} \Rightarrow$$

$$\Rightarrow P_{Tx} = \frac{I_S^2(\omega L_{Tx})^2}{2}\frac{1}{R_{Tx}(k_{Tx\text{-}Rx}^2 Q_{Tx}Q_{Rx\text{-}L}+1)} \left.\begin{array}{c} \\ (2.13) \end{array}\right\} \Rightarrow \qquad (2.20)$$

$$\Rightarrow P_{MN} = \eta_{Link}P_{Tx} = \frac{I_S^2(\omega L_{Tx})^2}{2R_{Tx}}\frac{Q_{Rx\text{-}L}}{Q_L}\frac{k_{Tx\text{-}Rx}^2 Q_{Tx}Q_{Rx\text{-}L}}{(k_{Tx\text{-}Rx}^2 Q_{Tx}Q_{Rx\text{-}L} + 1)^2}$$

References

1. M. Schormans, V. Valente, A. Demosthenous, Practical inductive link design for biomedical wireless power transfer: a tutorial. IEEE Trans. Biomed. Circuits Syst. **12**(5), 1112–1130 (2018). ISSN 1932-4545. https://doi.org/10.1109/TBCAS.2018.2846020

2. K. Van Schuylenbergh, R. Puers, *Inductive Powering: Basic Theory and Application to Biomedical Systems* (Springer, Netherlands, 2009)

3. L. Chen, S. Liu, Y.C. Zhou, T.J. Cui, An optimizable circuit structure for high-efficiency wireless power transfer. IEEE Trans. Ind. Electron. **60**(1), 339–349 (2013). ISSN 0278-0046. https://doi.org/10.1109/TIE.2011.2179275

4. R.F. Xue, K.W. Cheng, M. Je, High-efficiency wireless power transfer for biomedical implants by optimal resonant load transformation. IEEE Trans. Circuits Syst. I **60**(4), 867–874 (2013). ISSN 1549-8328. https://doi.org/10.1109/TCSI.2012.2209297

5. M. Kiani, U.M. Jow, M. Ghovanloo, Design and optimization of a 3-coil inductive link for efficient wireless power transmission. IEEE Trans. Biomed. Circuits Syst. **5**(6), 579–591 (2011). ISSN 1932-4545. https://doi.org/10.1109/TBCAS.2011.2158431

6. M. Kiani, M. Ghovanloo, The circuit theory behind coupled-mode magnetic resonance-based wireless power transmission. IEEE Trans. Circuits Syst. I **59**(9), 2065–2074 (2012). ISSN 1549-8328

7. M. Kiani, B. Lee, P. Yeon, M. Ghovanloo, A Q-modulation technique for efficient inductive power transmission. IEEE J. Solid-State Circuits **50**(12), 2839–2848 (2015). ISSN 0018-9200. https://doi.org/10.1109/JSSC.2015.2453201

8. Y.H. Sohn, B.H. Choi, E.S. Lee, G.C. Lim, G. Cho, C.T. Rim, General unified analyses of two-capacitor inductive power transfer systems: equivalence of current-source SS and SP compensations. IEEE Trans. Power Electron. **30**(11), 6030–6045 (2015). ISSN 0885-8993. https://doi.org/10.1109/TPEL.2015.2409734

9. H. Kim, C. Song, D.H. Kim, D.H. Jung, I.M. Kim, Y.I. Kim, J. Kim, S. Ahn, J. Kim, Coil design and measurements of automotive magnetic resonant wireless charging system for high-efficiency and low magnetic field leakage. IEEE Trans. Microw. Theory Techn. **64**(2), 383–400 (2016). ISSN 0018-9480. https://doi.org/10.1109/TMTT.2015.2513394

10. P. Pérez-Nicoli, A. Rodríguez-Esteva, F. Silveira, Bidirectional analysis and design of RFID using an additional resonant coil to enhance read range. IEEE Trans. Microw. Theory Techn. **64**(7), 2357–2367 (2016). ISSN 0018-9480. https://doi.org/10.1109/TMTT.2016.2573275

11. R.R. Harrison, Designing efficient inductive power links for implantable devices, in *IEEE International Symposium on Circuits and Systems* (2007), pp. 2080–2083. https://doi.org/10.1109/ISCAS.2007.378508

12. M. Soma, D.C. Galbraith, R.L. White, Radio-frequency coils in implantable devices: misalignment analysis and design procedure. IEEE Trans. Biomed. Eng. **BME-34**(4), 276–282 (1987). ISSN 0018-9294. https://doi.org/10.1109/TBME.1987.326088

13. J.A. Ferreira, Improved analytical modeling of conductive losses in magnetic components. IEEE Trans. Power Electron. **9**(1), 127–131 (1994). ISSN 0885-8993. https://doi.org/10.1109/63.285503

14. F.W. Grover, *Inductance Calculations: Working Formulas and Tables.* Courier Corporation (2004)

15. S.S. Mohan, M. del Mar Hershenson, S.P. Boyd, T.H. Lee, Simple accurate expressions for planar spiral inductances. IEEE J. Solid-State Circuits **34**(10), 1419–1424 (1999)

16. H.A. Wheeler, Formulas for the skin effect. Proc. IRE **30**(9), 412–424 (1942)

17. U.-M. Jow, M. Ghovanloo, Modeling and optimization of printed spiral coils in air, saline, and muscle tissue environments. IEEE Trans. Biomed. Circuits Syst. **3**(5), 339–347 (2009)

18. U.-M. Jow, M. Ghovanloo, Design and optimization of printed spiral coils for efficient transcutaneous inductive power transmission. IEEE Trans. Biomed. Circuits Syst. **1**(3), 193–202 (2007)

19. F. Jolani, Y. Yu, Z. Chen, A planar magnetically coupled resonant wireless power transfer system using printed spiral coils. IEEE Antennas Wirel. Propag. Lett. **13**, 1648–1651 (2014)
20. S. Gevorgian, H. Berg, H. Jacobsson, T. Lewin, Application notes-basic parameters of coplanar-strip waveguides on multilayer dielectric/semiconductor substrates, part 1: high permittivity superstrates. IEEE Microw. Mag. **4**(2), 60–70 (2003)
21. S. Raju, R. Wu, M. Chan, C.P. Yue, Modeling of mutual coupling between planar inductors in wireless power applications. IEEE Trans. Power Electron. **29**(1), 481–490 (2013)
22. A.P. French, *Vibrations and Waves* (CRC Press, 2017)
23. H.A. Haus, *Waves and Fields in Optoelectronics* (Prentice-Hall, 1984)
24. Y. Zhang, Z. Zhao, Frequency splitting analysis of two-coil resonant wireless power transfer. IEEE Antennas Wirel. Propag. Lett. **13**, 400–402 (2014). ISSN 1536-1225. https://doi.org/10.1109/LAWP.2014.2307924
25. A.P. Sample, D.T. Meyer, J.R. Smith, Analysis, experimental results, and range adaptation of magnetically coupled resonators for wireless power transfer. IEEE Trans. Ind. Electron. **58**(2), 544–554 (2011). ISSN 0278-0046. https://doi.org/10.1109/TIE.2010.2046002
26. Y. Lv, F. Meng, B. Che, Y. Wu, Q. Wu, L. Sun, A novel method for frequency splitting suppression in wireless power transfer, in *IEEE Asia-Pacific Conference on Antennas and Propagation* (2014), pp. 590–592. https://doi.org/10.1109/APCAP.2014.6992563
27. Y. Lyu, F. Meng, G. Yang, B. Che, Q. Wu, L. Sun, D. Erni, J.L. Li, A method of using nonidentical resonant coils for frequency splitting elimination in wireless power transfer. IEEE Trans. Power Electron. **30**(11), 6097–6107 (2015). ISSN 0885-8993. https://doi.org/10.1109/TPEL.2014.2387835
28. L. Jianyu, T. Houjun, G. Xin, Frequency splitting analysis of wireless power transfer system based on T-type transformer model. Elektronika Elektrotechnika **19**(10), 109–113 (2013)

Chapter 3
Inductive Link: Practical Aspects

3.1 Coil Design

In Chap. 2, the inductive link performance was attributed to its Power Transfer Efficiency (PTE), η_{Link}, and the power delivered to the Rx-circuit, P_{MN}. These parameters and most analytical expressions deduced in Chap. 2 depend on the coils' quality factors, Q_i, and coupling coefficients, k_{ij}. Particularly, in Sect. 2.1.5, the effects of Q_{Tx}, Q_{Rx}, and $k_{Tx\text{-}Rx}$ in a 2-coil link were analyzed and quantified in an example. Not only coils' electrical but also geometrical design as well as their relative spatial arrangement determine the Q_i and k_{ij}, which are addressed in this chapter in more detail.

On one hand, to maximize the link efficiency, η_{Link}, the coils should be designed and positioned to have the maximum possible Q_{Tx}, Q_{Rx}, and $k_{Tx\text{-}Rx}$. On the other hand, increasing Q_{Tx}, Q_{Rx}, and $k_{Tx\text{-}Rx}$ without considering Tx-Rx impedance mismatch may be detrimental and end up reducing P_{MN}. These points were initially discussed in Sect. 2.1.5. Increasing the quality factors beyond a certain limit may also negatively affect Intersymbol Interference (ISI) in back telemetry, if that is also expected from the inductive link, which will be discussed in Chap. 4. In low-frequency RFID, instead of maximizing the coils' Q-factor, even additional resistance may need to be added to reduce Q-factor and limit the ISI. Therefore, in design of inductive links, it is important to identify the desired range for Q-factor and k_{ij} depending on the application and target characteristics of the inductive WPT link.

As mentioned in Sect. 2.1.1, self-inductance is related to the voltage induced in the same conductor that carries a time-varying current. The geometrical disposition of the conductor (e.g., if it is coiled and how it is coiled) affects the magnetic field distribution that reaches each part of the conductor and defines the induced voltage. Furthermore, the induced voltage produces eddy currents in the conductor that alter the field and current distribution in the conductor cross-section. This originates the skin and proximity effects [1], which are discussed later in Sect. 3.1.1. These effects

P. Pérez-Nicoli et al., *Inductive Links for Wireless Power Transfer*,
https://doi.org/10.1007/978-3-030-65477-1_3

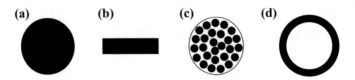

Fig. 3.1 Types of wires used in WPT coils. (**a**) Solid wire. (**b**) Printed wire. (**c**) Litz wire. (**d**) Hollow copper tubes

become more significant as the frequency is increased and, in many cases, are key factors in defining the value of the parasitic Equivalent Series Resistance (ESR). In this context, there are a few other engineering considerations, such as ease of manufacturing, weight, lifetime, cost, etc. that are often taken into account in mass production of WPT coils and result in several popular coil structures and conductor types, which are out of the scope of this chapter. Nonetheless, the type of coils used in inductive links can be categorized based on (1) the type of conductor used, e.g., printed circuit, solid (magnet) wire, litz wire, hollow copper tube, etc.; (2) how they are winded, e.g., planar-spiral or solenoid; (3) the shape of the coil, e.g., circular, rectangular, or hexagonal; and (4) the use of ferrite or other materials as cores, ferrite shielding, or other magnetic or insulating materials inside and around the coils to achieve the desired performance.

In Fig. 3.1 some examples of conductors used in WPT are presented. The cross-section of the wire can be circular, when a typical insulated wire is used (Fig. 3.1a), or rectangular, if the coil is printed on a rigid or flexible circuit board (Fig. 3.1b). The cross-section of the conductor is particularly important at higher carrier frequencies because of the skin and proximity effects [1–3], which result in the current density to be larger at certain points near the surface of the conductor, instead of being evenly distributed across the conductor cross-section. These effects generally increase the coil resistance and reduce its Q-factor. Therefore, depending on the carrier frequency, designers tend to choose conductor cross-sections that would maximize their surface area. For instance, to achieve higher coil Q-factor, litz wire can be used, as shown in Fig. 3.1c. The litz wire is a multi-strand bundle of wires individually insulated and woven together with the aim of reducing the ESR at high frequency by increasing the surface area of the conductor. The number and diameter of strands can be chosen in a way to further increase the coil Q-factor in the desired frequency range [4]. Above 1 MHz, it has been proven that hollow tubes (Fig. 3.1d) can achieve even lower resistance than litz wires at the cost of increased volume of the coil [5].

Examples of coil shapes can be found in Table 3.1, where the coils are divided between Printed Spiral Coils (PSC) and helical coils. The coil shape selection is mainly determined by the required geometry for the intended application. For instance, PSCs are easy to shape and batch fabricate at low cost through lithography and preferred when there is a constraint on the thickness of the device, e.g., a credit card, or when the coil needs to be flexible and conform to the surface of a curved object or human body. On the other hand, they impose additional limitations, such as increasing their number of turns without increasing their diameter resulting

Table 3.1 Different shapes of WPT coils

	Planar spiral			
	[6, 7] Printed wire Fig. 3.1b	[7–9] Solid wire Fig. 3.1a	[10] Hollow tube Fig. 3.1d	[11] Litz wire Fig. 3.1c
	[12–16] Printed wire single-layer Fig. 3.1b	[17, 18] Printed wire double-layer Fig. 3.1b	[19] Litz wire multi-layer Fig. 3.1c	
	Helical			
	[15, 20–23] Solid wire single-layer Fig. 3.1a	[5] Hollow tube single-layer Fig. 3.1d	[24] Hollow tube multi-layer Fig. 3.1d	[25, 26] Litz wire multi-layer Fig. 3.1c
	[27, 28] Litz wire single-layer Fig. 3.1c	[29] Litz wire multi-layer Fig. 3.1c		

in narrower conductive traces, higher ESR, and lower Q-factor. Unlike PSCs, helical coils cannot be batch fabricated at least cost-effectively and require special winding machines. However, they generally have fewer constraints and can offer considerably higher Q-factors.

As shown in earlier chapters, for theoretical calculations, each coil should be replaced with a lumped equivalent electrical model, whose level of complexity is a trade-off between accuracy and cost of computation. Figure 3.2 shows one of the simplest models that are commonly used to determine the link performance, in which the coil's distributed self-inductance, parasitic ESR, and parasitic capacitance between the turns are lumped in L, R, and C, respectively. It is difficult to find closed-form theoretical approximations that can estimate L, R, and C for every coil geometry. Instead, such empirical approximations and theoretical models are derived for particular coil shapes, dimensions, or types of wire either in closed-form or tabular format [2, 12, 30–35]. Additionally, the electrical model is affected by the presence of materials that affect the magnetic and/or electric fields (e.g., ferrites, metals, biological tissue) in the coil surroundings [18, 32, 36]. As a result, although theoretical models can be used for preliminary estimates and theoretical optimization of the coil geometry, software-aided design tools generally known as field solvers (see Table 2.4) are usually required for more accurate and complex designs.

In the following section, we present theoretical expressions to estimate the lumped model parameters in Fig. 3.2 for a square-shaped PSC.

Fig. 3.2 Coil equivalent
lumped electrical model

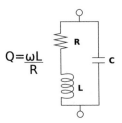

$$Q = \frac{\omega L}{R}$$

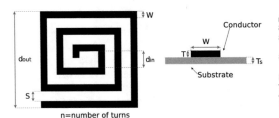

ρ = conductor resistivity
ε_0 = free space electric permittivity
ε_{rt} = relative permittivity of volume conductor
    (e.g. air)
ε_{rs} = relative permittivity of substrate
    (e.g. FR4)
μ_0 = free space magnetic permeability

The permeability of all other materials, such as air,
copper or substrate (FR4) are approximated by μ_0
ω = angular frequency

Fig. 3.3 Parameters that define the square-shaped PSC

3.1.1   Square-Shaped Printed Spiral Coil

The geometrical parameters that define a square-shaped PSC are presented in
Fig. 3.3, which are subsequently used in estimating the self-inductance, ESR,
parasitic capacitance , and mutual inductance. We present here a summary of these
estimations, their deductions are out of the scope of this section, and they can be
found in the corresponding references. At the end of this section, these parameters
are derived numerically in a practical example.

3.1.1.1   Self-Inductance, L

A simple expression was derived in [31], which can be applied with different coefficients to estimate the self-inductance of PSCs with square, hexagonal, octagonal,
and circular shapes, which is presented in (3.1). The parameters c_1, c_2, c_3, and c_4
depend on the shape of the coil, n is the number of turns, and d_{out} and d_{in} are the
external and internal dimensions depicted in Fig. 3.3. For the square-shaped coil
addressed in this section, the parameters c_1, c_2, c_3, and c_4, are the ones indicated in
(3.1) [31]:

$$L = \frac{\overbrace{c_1}^{1.27} \mu_0 n^2 d_{avg}}{2} \left[\ln \left(\frac{\overbrace{c_2}^{2.07}}{\varphi} \right) + \overbrace{c_3}^{0.18} \varphi + \overbrace{c_4}^{0.13} \varphi^2 \right]$$

$$d_{avg} = \frac{1}{2}(d_{out} + d_{in})$$

$$\varphi = \frac{d_{out} - d_{in}}{d_{out} + d_{in}}$$

(3.1)

The accuracy of (3.1) degrades as the ratio S/W becomes larger, where S is the turn spacing and W is the turn width, it exhibits a maximum error of 8% for $S \leq 3W$. The turn spacing could be enlarged to reduce interwinding capacitance; however, typically in PSCs for WPT, $S \leq W$ to reduce the area used by the spiral.

3.1.1.2 Equivalent Series Resistance (ESR)

Even though calculating the DC resistance of a conductor is straightforward once the geometry and resistivity of the conductor is known, estimating its AC resistance, which is a function of material, geometry, and frequency, is complex. Among numerous other factors, two important phenomena that affect the AC resistance of a conductor in the context of a PSC are the skin effect and proximity effect [1] among the adjacent turns of the PSC. The skin effect models the current distribution across the cross-section of the conductor, which tends to be larger near the surface. The proximity effect models the influence of the magnetic field generated by every turn of the inductor on the current passing through other turns, particularly the ones that are closest to that turn.

Formulas to estimate the ESR considering the skin effect were deduced in [2] for different conductor shapes. A simplified approximation valid for printed spiral coils is presented in [32]:

$$R_{skin} = \rho \underbrace{\frac{l}{W.T}}_{R_{DC}} \frac{T}{\delta(1 - e^{-T/\delta})} \frac{1}{(1 + T/W)},$$

(3.2)

where δ is the skin depth

$$\delta = \sqrt{\frac{2\rho}{\mu_0 \omega}},$$

(3.3)

ρ is the conductor resistivity, and l is the length of the wire which can be calculated from the coil dimensions as [33]

$$l = 4.n.d_{\text{out}} - 4.n.W - (2n + 1)^2(S + W). \tag{3.4}$$

The other parameters, W, S, T, d_{out}, and n, were defined in Fig. 3.3.

The proximity effect further increases the coil resistance. The rough estimate in (3.5) has been proposed in [3] to model the extra resistance due to proximity effects in a PSC. The ω_{crit} represents the frequency at which the proximity effect becomes significant [3]:

$$R_{prox} = \frac{R_{DC}}{10} \left(\frac{\omega}{\omega_{crit}} \right)^2 \tag{3.5}$$

$$\omega_{crit} = \frac{3.1}{\mu_0} \frac{(S + W)}{W^2} \frac{\rho}{T} \tag{3.6}$$

The total ESR, R_{TOT}, considering the skin and the proximity effect, can be estimated as the sum of R_{skin} (3.2) and R_{prox} (3.5) [32], as

$$R_{TOT} = \rho \underbrace{\frac{l}{W.T}}_{R_{DC}} \left[\frac{T}{\delta(1 - e^{-T/\delta})} \frac{1}{1 + T/W} + \frac{1}{10} \left(\frac{\omega}{\omega_{crit}} \right)^2 \right]. \tag{3.7}$$

More accurate estimates of the proximity effect have since been proposed, such as [12]

$$R_{prox} = R_{DC}(\phi f + \psi f^2) \tag{3.8}$$

where ϕ and ψ are fitting coefficients that are empirically determined and f is the carrier frequency. The need for using empirically defined parameters highlights the difficulty in predicting the proximity effects using closed-form mathematical equations.

3.1.1.3   Parasitic Capacitance, C

A simplified approximation of the parasitic capacitance of an inductor can be derived using the parallel plate capacitance formulation, which considers the electrical permittivity of the space between the conductive plates multiplied by the ratio between the plates' area and separation $\left(C = \varepsilon_0 \frac{\text{plates area}}{\text{plates separation}} \right)$. In a PSC, part of the electric field passes through the air or other environment surrounding the PSC (e.g., tissue in the case of implantable devices), and another part passes through the substrate (e.g., FR4 or polyimide printed circuit boards). Therefore, the difference between permittivities of air and FR4 should be considered in this approximation. In [33], the parasitic capacitance was estimated as

$$C = (\underbrace{\alpha}_{\simeq 0.9} \times \overbrace{\varepsilon_{rt}}^{\simeq 1} + \underbrace{\beta}_{\simeq 0.1} \times \overbrace{\varepsilon_{rs}}^{\simeq 4.4})\varepsilon_0 \frac{T}{S} l_g, \tag{3.9}$$

where α and β are empirically determined parameters; ε_{rt} and ε_{rs} are the relative dielectric constants of the surrounding (e.g., air) and substrate (e.g., FR4) materials, respectively; and l_g is the gap length between conductors, which is slightly shorter than l (wire length), and can be calculated as

$$l_g = 4(d_{out} - W.n)(n - 1) - 4S.n(n + 1). \tag{3.10}$$

The parameters T, S, W, d_{out} and n, used in (3.9) and (3.10), were defined in Fig. 3.3.

The expression presented in (3.9) is a rough estimation specially when the turns are far apart compared to their thickness, a condition in which edge effects dominate and the capacitance cannot be simply estimated by the two parallel-plate models. A more accurate but complex estimation for this parasitic capacitance is presented in [34]:

$$C = \varepsilon_0 \varepsilon_{eff} \frac{K(k_0')}{K(k_0)} l \tag{3.11}$$

$$\varepsilon_{eff} = 1 + (\varepsilon_{rs} - 1)\frac{1}{2} \frac{K(k_1')}{K(k_1)} \frac{K(k_0)}{K(k_0')} \ ; \quad k_0 = \frac{S}{S+2W} \ ; \quad (3.4) \ ; \quad k_0' = \sqrt{1 - k_0^2}$$

$$k_1 = \frac{\tanh\left(\frac{\pi S}{4T_S}\right)}{\tanh\left(\frac{\pi(S+2W)}{4T_S}\right)} \qquad k_1' = \sqrt{1 - k_1^2}$$

where $K(\cdot)$ represents the complete elliptic integral of the first kind and the other parameters, S, W, ε_0, and ε_{rs} were defined in Fig. 3.3.

It can be noted that the copper thickness, T, was not considered in (3.11). This is because a copper layer width is much higher than its thickness ($W \gg T$). In [32] it is shown that if this is not the case, the copper thickness can be considered as an increment in its width, as presented in Fig. 3.4 and (3.12).

$$\Delta = \frac{T}{2\pi \varepsilon_e} \left[1 + \ln\left(\frac{8\pi W}{T}\right) \right] \tag{3.12}$$

where ε_e is the mean value of permittivity of the layers in contact with the strips. For the case of PSCs, considering $T = 35 \, \mu m$, $W = 1 \, mm$, and $\varepsilon_e = 4$, $\Delta \simeq 11 \, \mu m$ can be neglected compared with $W = 1 \, mm$.

A significant consequence of the parasitic capacitance of the coil is that at a high enough frequency, the impedance of the coil becomes capacitive (i.e., dominated by the capacitance in the model of Fig. 3.2) instead of inductive. This occurs at

Fig. 3.4 Conductor width
adaptation to consider the
effect of the conductor
thickness in the parasitic
capacitance

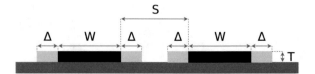

frequencies above the Self-Resonant Frequency (SRF) of the inductor that is given
by (3.13) and corresponds to the frequency where the inductance and capacitance
of Fig. 3.2 resonate. In (3.13), L and C are the self-inductance and parasitic
capacitance that were previously presented in the coil equivalent electrical model
(Fig. 3.2) and calculated in the previous subsections:

$$f_{self} = \frac{1}{2\pi\sqrt{LC}}.$$
(3.13)

3.1.1.4 Mutual Inductance, M

A compact model of mutual inductance between two planar inductors was proposed
in [35], which was extended in [12]. In this approach, the turns of the PSC are
modeled as constant current-carrying filaments, and the mutual inductance between
individual filaments is determined by solving Neumann's formula [30], using a
series expansion technique. Using this model, for two planar square-shaped coils, Tx
and Rx, assumed to be parallel and co-centric, the $M_{Tx\text{-}Rx}$ can be approximated as

$$M_{Tx\text{-}Rx} = \left(\frac{4}{\pi}\right)^{\left(1+\frac{d_{out_{Rx}}}{d_{out_{Tx}}}\right)} \sum_{i=1}^{i=n_{Tx}} \sum_{j=1}^{j=n_{Rx}} M_{ij}$$

$$M_{ij} = \frac{\mu_0 \pi a_i^2 b_j^2}{2(a_i^2 + b_j^2 + D_{Tx\text{-}Rx}^2)^{3/2}} \left(1 + \frac{15}{32}\gamma_{ij}^2 + \frac{315}{1024}\gamma_{ij}^4\right)$$

$$a_i = \frac{d_{out_{Tx}}}{2} - (i-1)(W_{Tx} + S_{Tx}) - \frac{W_{Tx}}{2}$$ (3.14)

$$b_j = \frac{d_{out_{Rx}}}{2} - (j-1)(W_{Rx} + S_{Rx}) - \frac{W_{Rx}}{2}$$

$$\gamma_{ij} = \frac{2a_i b_j}{a_i^2 + b_j^2 + D_{Tx\text{-}Rx}^2},$$

where the subscripts Tx and Rx were added to indicate if the parameters (W, S, n
or d_{out}, defined in Fig. 3.3) correspond to the Tx or Rx coil. In (3.14) it was also
assumed that $d_{out_{Tx}} \geq d_{out_{Rx}}$.

Since the turns are modeled as filaments, this approximation is only valid to
calculate the mutual inductance between coils that have the turn width, W, much
less than the diameter.

The coupling coefficient, $k_{Tx\text{-}Rx}$, can be calculated as

$$k_{Tx\text{-}Rx} = M_{Tx\text{-}Rx}/\sqrt{L_{Tx}L_{Rx}} \qquad (3.15)$$

The coupling coefficient could be directly estimated, instead of calculating it from the self and mutual inductance. Although approximations for the $k_{Tx\text{-}Rx}$ of square-shaped coils are not typical, a well-known expression for circular coils was proposed by Roz and Fuentes [37]. That estimation is presented in (3.16) where r_{Tx} and r_{Rx} are the Tx and Rx coils radius, respectively, and $D_{Tx\text{-}Rx}$ is the distance between them, and it is assumed that the coils are coplanar, perfectly aligned, and $r_{Tx} \geq r_{Rx}$:

$$k_{Tx\text{-}Rx} = \frac{r_{Tx}{}^2 . r_{Rx}{}^2}{\sqrt{r_{Tx}.r_{Rx}}\left(\sqrt{D_{Tx\text{-}Rx}{}^2 + r_{Tx}{}^2}\right)^3} \qquad (3.16)$$

From (3.16), it can be seen that the coupling coefficient, contrary to the self and mutual inductance, does not depend on the number of turns. The estimation of (3.16) is used throughout the book in many examples.

3.1.1.5 Square-Shaped Printed Spiral Coil Example

The previously presented expressions are utilized in an example in Table 3.2 and compared with simulation results using two different field solver software, CST and Sim4Life.

3.2 Influence of Foreign Object

The electrical model of the coil is affected by the presence of any conductor, dielectric, or ferromagnetic material near the coil, and this influence is stronger the larger or the closer the object is. For instance, the Rx-circuit should be considered as a nearby object that can potentially affect the Rx coil performance, particularly if it has large areas of conductive materials (such as ground planes) or other materials that can act as a magnetic shield. AIMDs intended for long-term implantation are usually enclosed inside a titanium case, which is a conductive material, and may affect the WPT link performance if it is not considered in the design [38]. Human tissue, blood, and other fluids also affect the link. At lower frequencies, biological tissue can be considered to have a magnetic permeability that is similar to the free space; however, the electric permittivity is more variable and particularly dependent on the type of tissue [39]. Therefore, in the presence of human tissue, the parasitic capacitance is more affected than the self-inductance. Ferrites are also used for the

Table 3.2 Square-shaped PSC example. The calculation script is available in the supplementary material, file Sec311

Constants		
Free space magnetic permeability	μ_0	$4\pi \times 10^{-7}$ H/m
Free space electric permittivity	ε_0	8.85×10^{-12} F/m
Coil parameters		
Conductor width	W	1 mm
Conductor spacing	S	3 mm
Conductor thickness	T	35 μm
Inner diameter	d_{in}	5 mm
Number of turns	n	5
Conductor resistivity	ρ	17 nΩm
Carrier frequency	f	13.56 MHz
Link parameters		
Identical parallel coils and perfect alignment is assumed		
distance between coils	$D_{Tx\text{-}Rx}$ 1 cm	

Estimations			
Parameter	Calculated	CST	Sim4Life
L	528 nH	570 nH	570 nH
R	0.29 Ω	0.32 Ω	0.32 Ω
C	5.2 pF	0.85 pF	n/a
M	193 nH	161 nH	159 nH

electromagnetic shield, enhancing magnetic coupling and protecting the users or other nearby instruments.

The effects of these objects that surround the coils can be included in the theoretical expressions of the equivalent electrical model, as the ones deduced in Sect. 3.1.1. For instance, [32] addresses the effect of coating and human tissue that surrounds the coil in estimating the parasitic capacitance. Expressions for mutual inductance and self-inductance with ferrite cores are presented in [40]. Including the effects of these objects on the inductive link would make the theoretical approximations even more complex to the extent that they end up being valid only for very specific conditions. Therefore, the use of computer-aided design is even more common when these objects and their effects are present in the design.

In the following sections, we present an overview of the effects of conductors and ferrites.

3.2.1   Effects of Conductive Materials

When a conductive material is placed near the inductive link, eddy currents are induced in it, affecting the magnetic field distribution and increasing power losses. As mentioned earlier, it is difficult to take into account these effects in theoretical

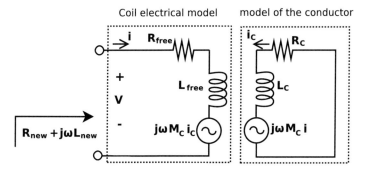

Fig. 3.5 Equivalent circuit of the conductor modeled as another coupled coil that is placed near the original coil

approximations. Therefore, this is usually addressed using Finite Element Analysis (FEA), implemented in field solver simulation software.

A simple approach to address the presence of a conductor near the link is to consider it as an extra, non-resonant, coupled coil [36, 41]. Let us consider a coil that, in free space, has an $ESR = R_{free}$, self-inductance=L_{free}, and, for simplicity sake, a negligible parasitic capacitance. Figure 3.5 shows the equivalent circuit of the conductor modeled as another coupled coil that is placed near the original coil. R_C and L_C are unknown parameters that represent the resistance and self-inductance of the conductive material, respectively. The mutual inductance M_C models the coupling between the coil and the conductor, and it is also an unknown parameter. A new model for the coil can therefore be obtained by considering the presence of the conductor and calculating R_{new} and L_{new}, which consider the reflected impedance from the conductive material near the coil, as shown in (3.17).

$$\left. \begin{array}{l} V = (R_{free} + j\omega L_{free})i + j\omega M_C i_C \\ i_C = \frac{-j\omega M_C i}{j\omega L_C + R_C} \end{array} \right\} \Longrightarrow$$

$$R_{new} + j\omega L_{new} = \frac{V}{i} = \frac{1}{i}\left((R_{free} + j\omega L_{free})i - j\omega M_C \frac{j\omega M_C i}{j\omega L_C + R_C} \right) \Longrightarrow$$

$$= R_{free} + j\omega L_{free} + \frac{(\omega M_C)^2}{(\omega L_C)^2 + R_C^2}(R_C - j\omega L_C)$$

$$= \underbrace{R_{free} + \frac{(\omega M_C)^2}{(\omega L_C)^2 + R_C^2}R_C}_{R_{new}} + j\omega \underbrace{\left(L_{free} - \frac{(\omega M_C)^2}{(\omega L_C)^2 + R_C^2}L_C \right)}_{L_{new}}$$

$$(3.17)$$

From L_{new} in (3.17), it can be seen that the original self-inductance is reduced, and this reduction is the more significant the lower the resistivity of the conductive material is or the stronger the mutual inductance is. This is an intuitive result since in a non-conductive material, $R_C \longrightarrow \infty$, or when the conductive object is too far, then $M_C \longrightarrow 0$, which leads to the presence of the object not affecting the self-inductance of the coil.

Regarding the ESR, R_{new} (3.17), it is increased due to the extra losses in the conductor. Again, the closer to the coil the conductor is, the larger M_C is, and the higher the ESR is. However, the effect of the conductor resistivity in the ESR may not be easy to understand. The R_{new}, as a function of R_C, has a maximum value when $R_C = \omega L_C$. To understand this relationship, let us analyze the two extreme conditions of $R_C \longrightarrow 0$ and $R_C \longrightarrow \infty$. When $R_C \longrightarrow \infty$, $R_{new} = R_{free}$, because no current is induced in the conductive object. When $R_C \longrightarrow 0$, again $R_{new} = R_{free}$; large current is induced in the conductive object, but this results in no power dissipation in the material. When $R_C = \omega L_C$, on the other hand, R_{new} is at its maximum, and in that situation the power dissipated in the conductive object is maximized.

The model presented in Fig. 3.5 can be compared with the 2-coil link where the receiver, in this case the conducting object, is a non-resonant receiver coil. Then, the existence of this worst-case R_C, which absorbs maximum power maximizing the ESR (R_{new}), can be compared with the existence of a load resistance, R_L, which maximizes the power delivered to the load, as shown, for instance, in Fig. 2.14. The analysis of that load, R_L, which maximizes η_{Link}, is addressed in Chap. 5.

The existence of a worst-case resistivity which strongly affects the ESR, and thus the coil quality factor, was addressed in [36] for the case of an AIMD. AIMDs intended for long-term implantation are usually enclosed inside a titanium case which is a conductive material. The Rx coil should inevitably be close to this titanium case to keep the AIMD small, thus affecting the Rx coil and consequently the inductive link electrical model. Different titanium grades (alloys) exist with different resistivities, which effects are studied in [36].

Although the approach presented in this section may not be enough to quantify the influence of the conductive material in an inductive link, it is useful in understanding the effect and deducing conclusions about how to reduce such undesired effects.

To summarize, the nearby conductive materials negatively affect the coil, reducing its self-inductance and increasing its ESR, thus reducing its Q-factor and consequently the efficiency of the inductive link. When it is not possible to avoid having a conductive object near the receiver or transmitter, for instance, when the receiver coil is in proximity of the titanium casing of an AIMD, ferrites can be used to shield the inductive link and reduce the negative effect of the conductor, as will be discussed in the following section.

3.2.2   Effect of Ferrites

Ferrites can be utilized to reduce the negative effects of conductive materials near the inductive link [42]. Additionally, in large power links, such as in electric vehicles, shielding is required not only to improve the link efficiency but also for safety and Electromagnetic Compatibility (EMC) reasons, as presented in Sect. 3.3.

Including the effects of a ferrite in the estimation of a WPT performance, such as the ones presented in Sect. 3.1.1, is complicated, and the results would be difficult to generalize. Hence, in this section, we present simulation results to illustrate the effects of a ferrite.

Let us consider the three examples given in Table 3.3 where the self-inductances, ESR, coil quality factor, mutual inductance, and coupling coefficient, are presented for each situation. The first one is the same as Table 3.2, which is used as a benchmark. In the second example, a ground plane was included near the Rx coil. This conductor strongly affects the magnetic field distribution, reducing the self-inductance, especially L_{Rx}, and the mutual inductance $M_{Tx\text{-}Rx}$. Additionally, the ESR is slightly increased because extra power is dissipated in the ground plane due to the induced current in it. As a consequence, the coils' quality factor are reduced, deteriorating the link efficiency. In the third example, a ferrite layer with $\mu_r = 130$ is included to the outer surfaces of the coil substrates as illustrated in the cross-section figure. The ferrite redirects most of the magnetic flux to avoid the ground plane, as a shield. On the other hand, the ferrite introduces new losses, similar to the core in a transformer, and therefore, its saturation should be avoided.

Table 3.3 Examples to quantify the effects of ferrites in the link (simulation results using CST)

Coil parameters for both Tx and Rx: see Table 3.2			
Ferrite relative magnetic permeability $\mu_r = 130$			
f	13.56 MHz	13.56 MHz	13.56 MHz
L_{Tx}	570 nH	518 nH	685 nH
L_{Rx}	570 nH	254 nH	628 nH
R_{Tx}	0.32 Ω	0.33 Ω	0.36 Ω
R_{Rx}	0.32 Ω	0.36 Ω	0.38 Ω
Q_{Tx}	152	134	162
Q_{Rx}	152	60	141
$M_{Tx\text{-}Rx}$	161 nH	44 nH	195 nH
$k_{Tx\text{-}Rx}$	0.28	0.12	0.30

Even when shielding is not required, a ferrite can be used to improve the self- and mutual inductance, concentrating the magnetic flux. The use of ferrite in the link should be evaluated considering the available space, costs, biocompatibility (in AIMDs), and the link performance.

3.3   Safety and Electromagnetic Compatibility Considerations

The power received in the Rx-circuit of a WPT link is equal to the transmitter power P_{Tx} times the link efficiency η_{Link} (2.16). η_{Link} decreases with decreasing coupling coefficient $k_{Tx\text{-}Rx}$ (2.13). The coupling coefficient strongly depends on the distance (and alignment) between Tx and Rx (Fig. 2.9). From (3.16) it can be seen that at the large distances, it decreases with the cube of the distance. Therefore, to increase distance at a given received power, the transmitted power must be increased to compensate for a decreased coupling coefficient and link efficiency. Alternatively, increasing the received power at a given distance also requires to increase the transmitted power. These two scenarios lead to increased electromagnetic fields in the surroundings of the Tx. This section discusses the two reasons that limit in practice the intensity of the electromagnetic fields that a WPT link can generate in its surroundings. First, these fields might disturb other electronic equipment, which must be prevented. Therefore, this limit is related to the Electromagnetic Compatibility (EMC) of the equipment using WPT. Second, and most critical, an excessive intensity of these fields might produce adverse effects on the health of people in the proximity of the link. This limit is, thus related, to the safety of the users.

In considering the reference sources of information for these two limits, two kinds of sources must be identified: regulatory bodies and standards. It is useful to take into account their roles and characteristics.

Regulatory bodies define, with force of law, the requirements so that a product can be used in a given market. Examples of these are the following. For Europe, the directives are issued by the European Union and corresponding national legislation. In the United States, relevant to WPT systems are the rules and regulations issued by the Federal Communications Commission (FCC) and by regulatory agencies of the particular application field (e.g., the Food and Drug Administration (FDA) in the case of WPT applied to medical devices). As it is not possible to cover all the variety of cases of all national legislations worldwide, we will refer to the cases of Europe and the United States to illustrate what follows. Many other countries use regulations that are similar to these ones. Reference [43] does a comparison of the criteria of several regions and countries regarding the limits of human exposure to electromagnetic fields and [44] on the particular case of WPT. The reader should refer to the regulations of the particular country or region he/she is interested in.

Standards are usually developed by international, regional, or national organizations, based on the input from experts and industry. Some organizations that develop standards related to the topics of this section are as follows (non-exhaustive list): the Institute of Electrical and Electronic Engineers (IEEE), the International Commission on Non-Ionizing Radiation Protection (ICNIRP), the International Electrotechnical Commission (IEC), the European Telecommunications Standards Institute (ETSI), and the European Committee for Electrotechnical Standardization (CENELEC). Usually regulations are based on preexisting standards. Furthermore, standards are a key part of the process of demonstrating compliance with the regulations, since they specify in more detail than regulations how to test the compliance with the stated limits. The demonstration of compliance with the appropriate standards provides the presumption of conformity with the regulation. Nevertheless, the ultimate limits that need to be complied are those defined by the competent regulatory authority.

3.3.1 Electromagnetic Compatibility (EMC)

The requirements to assure compliance with EMC criteria is strongly based on the rules of allocation of the electromagnetic spectrum. Therefore, the exact limits are very dependent on the selected frequency for operation of the WPT link. And, conversely, the selection of this frequency, besides other technical considerations (e.g., the impact on the size of the coils), should take into account the limitations imposed by the EMC-related regulation that could relax the limits at certain frequencies or strongly limit, or even forbid, the operation in others. A detailed description of all cases is out of the scope of this book. We will point out some sources of information and exemplify some cases.

In the case of Europe, EMC and frequency use are regulated by Directive 2014/53/EU on radio communication equipment [45] and Directive 2014/30/EU on EMC [46] (applicable to equipment that is not radio communication equipment and therefore not covered by Directive 2014/53/EU). The limits are defined by the corresponding harmonized standards [47, 48]. In the case of the EMC directive, there are 162 harmonized standards dealing with different kinds of equipment. For example, CENELEC standard EN 55011 applies to industrial, scientific, and medical equipment [49].

In the United States, EMC in unlicensed bands are regulated in title 47 of the Code of Federal Regulations, Part 15 in the general case. In the particular case of Industrial, Scientific and Medical (ISM) purposes, not related to telecommunications, Part 18 applies. Depending on the operating configuration, wireless power transfer devices may need to be approved under Part 15, Part 18, or both [50].

Considering a WPT operating in the ISM band of 13.56 MHz, as established in 47 CFR Part 18, within this band (tolerance $+/- 7$ kHz, cl. 18.301), the equipment is permitted unlimited radiated energy in the band. On one hand, this imposes a requirement on the precision of the operating frequency. On the other hand, the actual limit for generated electromagnetic fields at the operating frequency will depend on the safety considerations that will be discussed next. At other frequencies outside of the band, 47 CFR Part 18, cl. 18.305 requires that the field strength is below 25 μV/m at a distance of 300 m. This imposes a constraint on the radiation at harmonics of the operating frequency, which depends on the harmonic content of the driver of the Tx coil and on the quality factor of the Tx resonant circuit.

3.3.2 Safety

The regulations and standards regarding safety criteria for exposure to electromagnetic fields are based on the analysis of several decades of published, scientifically substantiated, studies on potential adverse effects of electromagnetic fields on the human body. Current regulation and standards consider two hazards (or effects) [51–53]: tissue heating for frequencies above 100 kHz or nerve stimulation from contact or induced currents or fields in the body in the frequency range below 10 MHz.

Following is presented how the limits to keep these two effects within safe ranges are derived and expressed in the case of the last revision (2020) of the ICNIRP standard [51]. The IEEE Standard [52] follows a similar approach with slight differences.

It should be noted that for each case, four variants of the limits are stated based on the four combinations resulting from two possible situations for two aspects the exposure quantity considered (internal/external) and the exposure environment (occupational/general public in terms of ICNIRP [51] or restricted/unrestricted in terms of IEEE [52]).

The exposure limits are intrinsically dependent on the value of physical magnitudes (field intensity, energy absorption, or power density as we will later discuss) inside the body. When expressed in terms of this physical magnitudes inside the body, we have the internal exposure quantity, called basic restriction by ICNIRP [51] and dosimetric reference limit by IEEE [52]. However, actual measurements inside the body are not feasible. This is substituted by measurements on a phantom substance that represents the body and approximates its electromagnetic characteristics [53]. Particular procedures have been standardized for some cases (e.g., mobile phones), but not yet for WPT applications. The extended alternative to perform measurements on a phantom is to assess the required quantities through 3D numerical simulation using an anatomical model of the human body that can represent the dielectric properties of the different tissues (e.g., skin, fat, muscle, bones, brain, inner organs) [53].

Due to the difficulty to experimentally demonstrate compliance with the standards based on the quantities involved in the basic restrictions, the standards also state the limits in terms of quantities that are more easily assessed and correspond to the levels measured outside of the body. These are what are called reference levels by ICNIRP [51] and exposure reference level by IEEE [52]. These reference levels have been estimated so that they provide an equivalent level of protection to the basic restrictions for worst-case exposure scenarios. Thus, they will be more conservative than the basic restrictions in most cases. Furthermore, it has been shown that they are even more conservative in the particular case of WPT systems operating in the near field [53, 54].

Additionally, both internal and external limits depend on frequency and are based on time and spatial averaging criteria that will be later presented. The definition of these limits starts from the threshold (lowest exposure) required to cause harm (adverse health effect) and takes a safety margin or reduction factor from this threshold. The second aspect that leads to two sets of limits is the exposure environment, which results in different reduction factors. On one hand, there is the case of the general public in terms of ICNIRP [51] (unrestricted environment in terms of IEEE [52]) where a larger reduction factor is applied. On the other hand, it is the case of the occupationally exposed individuals or occupational limits in terms of ICNIRP [51] (restricted environment in terms of IEEE [52]) where the reduction factor is smaller than in the previous case. The rationale for using a smaller reduction factor in the case of occupationally exposed individuals is that these are defined as [51] "adults who are exposed under controlled conditions associated with their occupational duties, trained to be aware of potential radiofrequency EMF risks and to employ appropriate harm-mitigation measures, and who have the sensory and behavioral capacity for such awareness and harm-mitigation response." It is considered in the standard [51] that this group is "not deemed to be at greater risk than the general public, providing that appropriate screening and training is provided to account for all known risks." It is interesting to note that in both standards a pregnant woman is considered part of the general public in all cases (independently of her training). Having defined this general organization for the limits, we may proceed to consider which quantities are assessed. The basic restrictions regarding heating effects are stated as follows. From 100 kHz to 6 GHz, where electromagnetic fields penetrate deeply into the tissue, the Specific Absorption Rate (SAR), which is the power absorbed per unit mass (W/kg), is considered. This is averaged on the whole body when considering variations in the body core temperature and on a 10 g [51] or 1 g [52] mass when considering local heating effects. At frequencies above 6 GHz, where the energy is absorbed superficially at the skin level, the power density averaged over a small area (either $4\,cm^2$ or $1\,cm^2$) is considered instead of SAR [51, 52]. The corresponding reference levels [51] are stated in terms of incident E field, H field, power density, energy density, and induced electric current on a limb, depending on the frequency and averaging time considered. Finally, regarding lower frequency effects, such as nerve stimulation, from 100 kHz to 10 MHz, the quantity used to state the basic restriction is the induced electric field. The corresponding reference levels are defined with the peak incident E and H field strength. Table 3.4 [55] shows how the basic restrictions for the heating effect are derived in the last version of the ICNIRP standard [51], starting from the adverse health effect to be prevented and the corresponding threshold. Following the magnitudes included in Table 3.4 are defined:

Table 3.4 Derivation of basic restrictions in the case of heating effect (from 100 kHz to 300 GHz) in ICNIRP standard [51, 55]

Parameter	Frequency range	ΔT	Spatial averaging	Temporal averaging	Health effect level	Reduction factor	Workers	Reduction factor	General public
Core ΔT	100 kHz–300 GHz	1 °C	whole body average	30 min	4 W/kg	10	0.4 W/kg	50	0.08 W/kg
Local ΔT (head & torso)	100 kHz–6 GHz	2 °C	10 g	6 min	20 W/kg	2	10 W/kg	10	2 W/kg
Local ΔT (limbs)	100 kHz–6 GHz	5 °C	10 g	6 min	40 W/kg	2	20 W/kg	10	4 W/kg
Local ΔT (head & torso, limbs)	>6–300 GHz 30–300 GHz	5 °C	4 cm^2 1 cm^2	6 min 6 min	200 W/m^2 400 W/m^2	2	100 W/m^2 200 W/m^2	10	20 W/m^2 40 W/m^2

- **Parameter**: Health-related parameter to be limited, in all cases temperature elevation of different body parts: core of the body, head, torso, and/or limbs.
- **Frequency range**: Frequency range of the incident fields where the restriction is considered.
- **ΔT**: Maximum admissible temperature elevation.
- **Spatial averaging**: Mass or area considered for averaging the relevant quantity (SAR or power density)
- **Temporal averaging**: Time interval for taking the average of the relevant quantity.
- **Health effect level**: Threshold (lowest value) resulting on an adverse effect.
- **Reduction factor**: Factor that divides the health effect level to obtain the basic restriction.
- **Workers**: Basic restriction for occupationally exposed individuals.
- **General public**: Basic restriction for members of the general public.

It would be desirable to be able to translate in a general way the EMC and safety requirements in achievable performance (transmitted power, distance) of WPT systems. Some research efforts have been undertaken in this sense for safety requirements [56, 57]; nevertheless the results are still partial. Furthermore, studies show (e.g., [58]) important variability of the result not only with, as expected, the WPT system configuration and position with respect to the subject but also with the particular subject body geometry. Therefore, numerical simulation of the particular case is still mandatory. EMC criteria are expected to evolve to dedicated requirements for particular cases of WPT systems. The International Telecommunications Unit prepared a report [59] that recently summarized the state of standardization of WPT in the world, particularly from the EMC point of view.

An example of these trends in addressing EMC and safety is clear in the dynamic field of WPT for electric vehicles. On the EMC field, several normalization initiatives are being done [60], and some of them include also general design requirements [61–64]. The 85 kHz band appears as one of the established bands for this application. On the side of design for electromagnetic safety, several research efforts are ongoing: some of them have verified electromagnetic safety inside the vehicle cabin based on reference levels with a 7.7 kW wireless charger [65]. Others include verification of the basic restrictions as well [66]. Finally, it should be noted that most of these studies focus on the WPT link. However, EMC and safety requirements must be also analyzed regarding the WPT circuits (Tx and Rx).

References

1. H. Johnson, H.W. Johnson, M. Graham, *High-Speed Signal Propagation: Advanced Black Magic* (Prentice-Hall, 2003)
2. H.A. Wheeler, Formulas for the skin effect. Proc. IRE **30**(9), 412–424 (1942)
3. W.B. Kuhn, N.M. Ibrahim, Analysis of current crowding effects in multiturn spiral inductors. IEEE Trans. Microw. Theory Technol. **49**(1), 31–38 (2001)

4. C.R. Sullivan, Optimal choice for number of strands in a litz-wire transformer winding. IEEE Trans. Power Electron. **14**(2), 283–291 (1999)
5. S.H. Lee, R.D. Lorenz, Development and validation of model for 95%-efficiency 220-W wireless power transfer over a 30-cm air gap. IEEE Trans. Ind. Appl. **47**(6), 2495–2504 (2011)
6. X. Li, C.Y. Tsui, W.H. Ki, A 13.56 MHz wireless power transfer system with reconfigurable resonant regulating rectifier and wireless power control for implantable medical devices. IEEE J. Solid-State Circuits **50**(4), 978–989 (2015)
7. J. Kim, D.-H. Kim, Y.-J. Park, Analysis of capacitive impedance matching networks for simultaneous wireless power transfer to multiple devices. IEEE Trans. Ind. Electron. **62**(5), 2807–2813 (2014)
8. J. Kim, H.-C. Son, K.-H. Kim, Y.-J. Park, Efficiency analysis of magnetic resonance wireless power transfer with intermediate resonant coil. IEEE Antennas Wirel. Propag. Lett. **10**, 389–392 (2011)
9. D.H. Kim, J. Kim, Y.J. Park, Optimization and design of small circular coils in a magnetically coupled wireless power transfer system in the megahertz frequency. IEEE Trans. Microw. Theory Technol. **64**(8), 2652–2663 (2016) ISSN 0018-9480. https://doi.org/10.1109/TMTT.2016.2582874
10. Z. Pantic, B. Heacock, S. Lukic, Magnetic link optimization for wireless power transfer applications: modeling and experimental validation for resonant tubular coils, in *IEEE Energy Conversion Congress and Exposition* (IEEE, 2012), pp. 3825–3832
11. H. Kim, C. Song, D.H. Kim, D.H. Jung, I.M. Kim, Y.I. Kim, J. Kim, S. Ahn, J. Kim, Coil design and measurements of automotive magnetic resonant wireless charging system for high-efficiency and low magnetic field leakage. IEEE Trans. Microw. Theory Technol. **64**(2), 383–400 (2016). ISSN 0018-9480. https://doi.org/10.1109/TMTT.2015.2513394
12. F. Jolani, Y. Yu, Z. Chen, A planar magnetically coupled resonant wireless power transfer system using printed spiral coils. IEEE Antennas Wirel. Propag. Lett. **13**, 1648–1651 (2014)
13. M. Fu, T. Zhang, C. Ma, X. Zhu, Efficiency and optimal loads analysis for multiple-receiver wireless power transfer systems. IEEE Trans. Microw. Theory Technol. **63**(3), 801–812 (2015)
14. T.D. Yeo, D. Kwon, S.T. Khang, J.W. Yu, Design of maximum efficiency tracking control scheme for closed-loop wireless power charging system employing series resonant tank. IEEE Trans. Power Electron. **32**(1), 471–478 (2017)
15. C. Yang, C. Chang, S. Lee, S. Chang, L. Chiou, Efficient four-coil wireless power transfer for deep brain stimulation. IEEE Trans. Microw. Theory Technol. **65**(7), 2496–2507 (2017). ISSN 0018-9480. https://doi.org/10.1109/TMTT.2017.2658560
16. R.F. Xue, K.W. Cheng, M. Je, High-efficiency wireless power transfer for biomedical implants by optimal resonant load transformation. IEEE Trans. Circuits Syst. I **60**(4), 867–874 (2013). ISSN 1549-8328. https://doi.org/10.1109/TCSI.2012.2209297
17. K. Chen, Z. Zhao, Analysis of the double-layer printed spiral coil for wireless power transfer. IEEE J. Emerg. Sel. Top. Power Electron. **1**(2), 114–121 (2013)
18. M. Zargham, P.G. Gulak, Maximum achievable efficiency in near-field coupled power-transfer systems. IEEE Trans. Biomed. Circuits Syst. **6**(3), 228–245 (2012)
19. H. Li, J. Li, K. Wang, W. Chen, X. Yang, A maximum efficiency point tracking control scheme for wireless power transfer systems using magnetic resonant coupling. IEEE Trans. Power Electron. **30**(7), 3998–4008 (2015). ISSN 0885-8993. https://doi.org/10.1109/TPEL.2014.2349534
20. L. Chen, S. Liu, Y.C. Zhou, T.J. Cui, An optimizable circuit structure for high-efficiency wireless power transfer. IEEE Trans. Ind. Electron. **60**(1), 339–349 (2013). ISSN 0278-0046. https://doi.org/10.1109/TIE.2011.2179275
21. M. Fu, H. Yin, X. Zhu, C. Ma, Analysis and tracking of optimal load in wireless power transfer systems. IEEE Trans. Power Electron. **30**(7), 3952–3963 (2015)
22. T. Nagashima, X. Wei, E. Bou, E. Alarcón, M.K. Kazimierczuk, H. Sekiya, Steady-state analysis of isolated class-E^2 converter outside nominal operation. IEEE Trans. Ind. Electron. **64**(4), 3227–3238 (2017). ISSN 0278-0046. https://doi.org/10.1109/TIE.2016.2631439

23. Y. Cheng, G. Wang, M. Ghovanloo, Analytical modeling and optimization of small solenoid coils for millimeter-sized biomedical implants. IEEE Trans. Microw. Theory Technol. **65**(3), 1024–1035 (2016)

24. Z. Pantic, S. Lukic, Computationally-efficient, generalized expressions for the proximity-effect in multi-layer, multi-turn tubular coils for wireless power transfer systems. IEEE Trans. Magn. **49**(11), 5404–5416 (2013)

25. W.X. Zhong, S.Y.R. Hui, Maximum energy efficiency tracking for wireless power transfer systems. IEEE Trans. Power Electron. **30**(7), 4025–4034 (2015)

26. A.K. RamRakhyani, S. Mirabbasi, M. Chiao, Design and optimization of resonance-based efficient wireless power delivery systems for biomedical implants. IEEE Trans. Biomed. Circuits Syst. **5**(1), 48–63 (2010)

27. Q. Deng, J. Liu, D. Czarkowski, M.K. Kazimierczuk, M. Bojarski, H. Zhou, W. Hu, Frequency-dependent resistance of litz-wire square solenoid coils and quality factor optimization for wireless power transfer. IEEE Trans. Ind. Electron. **63**(5), 2825–2837 (2016)

28. L. Jinliang, D. Qijun, H. Wenshan, Z. Hong, Research on quality factor of the coils in wireless power transfer system based on magnetic coupling resonance, in *IEEE Workshop on Emerging Technologies, Wireless Power Transfer (WoW)* (IEEE, 2017), pp. 123–127

29. J. Shin, S. Shin, Y. Kim, S. Ahn, S. Lee, G. Jung, S.-J. Jeon, D.-H. Cho, Design and implementation of shaped magnetic-resonance-based wireless power transfer system for roadway-powered moving electric vehicles. IEEE Trans. Ind. Electron. **61**(3), 1179–1192 (2013)

30. F.W. Grover, *Inductance Calculations: Working Formulas and Tables* (Courier Corporation, 2004)

31. S.S. Mohan, M. del Mar Hershenson, S.P. Boyd, T.H. Lee, Simple accurate expressions for planar spiral inductances. IEEE J. Solid-State Circuits **34**(10), 1419–1424 (1999)

32. U.-M. Jow, M. Ghovanloo, Modeling and optimization of printed spiral coils in air, saline, and muscle tissue environments. IEEE Trans. Biomed. Circuits Syst. **3**(5), 339–347 (2009)

33. U.-M. Jow, M. Ghovanloo, Design and optimization of printed spiral coils for efficient transcutaneous inductive power transmission. IEEE Trans. Biomed. Circuits Syst. **1**(3), 193–202 (2007)

34. S. Gevorgian, H. Berg, H. Jacobsson, T. Lewin, Application notes-basic parameters of coplanar-strip waveguides on multilayer dielectric/semiconductor substrates, part 1: high permittivity superstrates. IEEE Microw. Mag. **4**(2), 60–70 (2003)

35. S. Raju, R. Wu, M. Chan, C.P. Yue, Modeling of mutual coupling between planar inductors in wireless power applications. IEEE Trans. Power Electron. **29**(1), 481–490 (2013)

36. P. Pérez-Nicoli, M. Biancheri-Astier, A. Diet, Y. Le Bihan, L. Pichon, F. Silveira, Influence of the titanium case used in implantable medical devices on the wireless power link, in *IEEE Wireless Power Transfer Conference* (IEEE, 2018), pp. 1–3

37. V. Fuentes, T. Roz, Using low power transponders and tags for RFID applications. EM Microelectron. Marin SA, Marin (1998)

38. C. Xiao, K. Wei, D. Cheng, Y. Liu, Wireless charging system considering eddy current in cardiac pacemaker shell: theoretical modeling, experiments, and safety simulations. IEEE Trans. Ind. Electron. **64**(5), 3978–3988 (2017)

39. M. Schormans, V. Valente, A. Demosthenous, Practical inductive link design for biomedical wireless power transfer: a tutorial. IEEE Trans. Biomed. Circuits Syst. **12**(5), 1112–1130 (2018) ISSN 1932-4545. https://doi.org/10.1109/TBCAS.2018.2846020

40. P.T. Theilmann, P.M. Asbeck, An analytical model for inductively coupled implantable biomedical devices with ferrite rods. IEEE Trans. Biomed. Circuits Syst. **3**(1), 43–52 (2008)

41. Y. Le Bihan, Study on the transformer equivalent circuit of eddy current nondestructive evaluation. NDT & E Int. **36**(5), 297–302 (2003)

42. T. Campi, S. Cruciani, F. Palandrani, V. De Santis, A. Hirata, M. Feliziani, Wireless power transfer charging system for AIMDs and pacemakers. IEEE Trans. Microw. Theory Technol. **64**(2), 633–642 (2016)

43. H.M. Madjar, Human radio frequency exposure limits: an update of reference levels in Europe, USA, Canada, China, Japan and Korea, in *IEEE International Symposium on Electromagnetic Compatibility* (IEEE, 2016), pp. 467–473
44. C. Kalialakis, A. Georgiadis, The regulatory framework for wireless power transfer systems. Wirel. Power Transf. **1**(2), 108–118 (2014)
45. European Parliament and the Council of the European Union, Directive 2014/53/eu of the European parliament and of the council. Official J. Eur. Union (2014). http://data.europa.eu/eli/dir/2014/53/oj. Accessed 11 June 2020
46. European Parliament and the Council of the European Union, Directive 2014/30/eu of the European parliament and of the council. Official J. Eur. Union (2014). http://data.europa.eu/eli/dir/2014/30/oj. Accessed 11 June 2020
47. The European Standardisation Organisations, Directive 2014/53/EU, harmonized standards, radio equipment (2016). https://ec.europa.eu/growth/single-market/european-standards/harmonised-standards/red_en. Accessed 11 June 2020
48. The European Standardisation Organisations, Directive 2014/30/EU, harmonized standards, electromagnetic compatibility (EMC) (2016). https://ec.europa.eu/growth/single-market/european-standards/harmonised-standards/electromagnetic-compatibility_en. Accessed 11 June 2020
49. CENELEC, Industrial, scientific and medical equipment – radio-frequency disturbance characteristics – limits and methods of measurement. Standard, CENELEC (2016)
50. Office of Engineering and Federal Communications Commission Technology Laboratory Division, RF exposure considerations for low power consumer wireless power transfer applications. Standard, Federal Communications Commission (2018)
51. ICNIRP, Guidelines for limiting exposure to electromagnetic fields (100 kHz to 300 GHz). Standard, International Commission on Non-Ionizing Radiation Protection, ICNIRP (2020)
52. IEEE Standards Coordinating Committee, IEEE standard for safety levels with respect to human exposure to electric, magnetic, and electromagnetic fields, 0 Hz to 300 GHz. *IEEE Std C95.1-2019 (Revision of IEEE Std C95.1-2005/Incorporates IEEE Std C95.1-2019/Cor 1-2019)* (2019), pp. 1–312. https://doi.org/10.1109/IEEESTD.2019.8859679
53. A. Christ, M. Douglas, J. Nadakuduti, N. Kuster, Assessing human exposure to electromagnetic fields from wireless power transmission systems. Proc. IEEE **101**(6), 1482–1493 (2013)
54. A. Christ, M.G. Douglas, J.M. Roman, E.B. Cooper, A.P. Sample, B.H. Waters, J.R. Smith, N. Kuster, Evaluation of wireless resonant power transfer systems with human electromagnetic exposure limits. IEEE Trans. Electromagn. Compat. **55**(2), 265–274 (2012)
55. ICNIRP, ICNIRP RF EMF guidelines in brief (2020). https://www.icnirp.org/en/publications/article/rf-guidelines-2020.html. Accessed 23 Sept 2020
56. K. Zhang, L. Du, Z. Zhu, B. Song, D. Xu, A normalization method of delimiting the electromagnetic hazard region of a wireless power transfer system. IEEE Trans. Electromagn. Compat. **60**(4), 829–839 (2017)
57. X.L. Chen, V. De Santis, A.E. Umenei, Theoretical assessment of the maximum obtainable power in wireless power transfer constrained by human body exposure limits in a typical room scenario. Phys. Med. Biol. **59**(13), 3453 (2014)
58. M. Koohestani, M. Ettorre, M. Zhadobov, Local dosimetry applied to wireless power transfer around 10 MHz: dependence on EM parameters and tissues morphology. IEEE J. Electromagn. RF Microw. Med. Biol. **2**(2), 123–130 (2018)
59. ITU, Wireless power transmission using technologies other than radio frequency beam. Report, International Telecommunication Union, ITU (2017). https://www.itu.int/pub/R-REP-SM.2303-2-2017
60. ETSI, Wireless power transmission systems, using technologies other than radio frequency beam in the 19–21 kHz, 59–61 kHz, 79–90 kHz, 100–300 kHz, 6765–6795 kHz ranges; harmonised standard covering the essential requirements of article 3.2 of directive 2014/53/EU. Standard, European Telecommunications Standards Institute, ETSI (2017)
61. J. Schneider, Wireless power transfer for light-duty plug-in/electric vehicles and alignment methodology. Standard, Society of Automotive Engineers, SAE (2019)

62. International Electrotechnical Commission et al., Electric vehicle wireless power transfer (WPT) systems – part 1: general requirements. Standard, International Electrotechnical Commission, IEC (2015)
63. International Electrotechnical Commission et al., Electric vehicle wireless power transfer (WPT) systems – part 2: specific requirements for communication between electric road vehicle (EV) and infrastructure. Technical specification, International Electrotechnical Commission, IEC (2019)
64. International Electrotechnical Commission et al., Electric vehicle wireless power transfer (WPT) systems – part 3: specific requirements for the magnetic field wireless power transfer systems. Technical specification, International Electrotechnical Commission, IEC (2019)
65. T. Campi, S. Cruciani, F. Maradei, M. Feliziani, Magnetic field during wireless charging in an electric vehicle according to standard SAE j2954. Energies **12**(9), 1795 (2019)
66. K. Miwa, T. Takenaka, A. Hirata, Electromagnetic dosimetry and compliance for wireless power transfer systems in vehicles. IEEE Trans. Electromagn. Compat. **61**(6), 2024–2030 (2019)

Chapter 4
Back Telemetry

4.1   The Need for and Role of Back Telemetry in WPT Links

In addition to the Wireless Power Transfer (WPT), there is often a need for wireless data transfer between the Tx and Rx. When data is transmitted in the same direction as the power, i.e., from the power transmitter, Tx, to the power receiver, Rx, it is referred to as forward telemetry. Whereas when the data is transmitted in the opposite direction compared to the power, i.e., from Rx to Tx, it is referred to as back telemetry, as illustrated in Fig. 4.1. Forward and back telemetry are also known as downlink and uplink, respectively.

Forward telemetry can be used to configure the Rx and/or send commands to request for information or control the Rx function, such as electrical stimulation of the neural tissue. Back telemetry can be used to obtain the Rx status, e.g., battery status, to collect the data measured by the Rx, e.g. neural recording, or to implement a feedback loop that controls the power transmission by opposing disturbances to the wireless link, such as changes in the coil distance or orientation, and loading variations. These feedback loops are addressed in detail in Chap. 7.

The data transmission can be implemented using a dedicated wireless link with its dedicated antennas, often at a higher frequency range. In that case, the data transmission link will face the same design challenges as any other Radio Frequency (RF) communication link with the additional challenge of strong interference from the power transmission link. In this approach, extra space is needed for housing the data transmission/reception antennas, and the dedicated data-Tx/Rx circuits will consume power [1, 2]. Therefore, in power-, size-, and cost-constrained systems, there is a tendency to use the same inductive link that is used for transferring power for data transmission [3–16].

This chapter is focused on implementing the back telemetry through the inductive power transfer link. In Sect. 4.2, the design of WPT links that need to support back telemetry is addressed. Then, in Sect. 4.3, two exemplar implementations are

P. Pérez-Nicoli et al., *Inductive Links for Wireless Power Transfer*,
https://doi.org/10.1007/978-3-030-65477-1_4

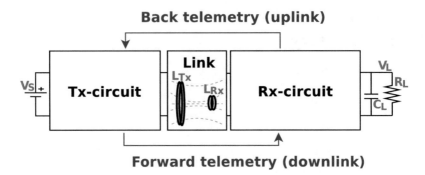

Fig. 4.1 Forward and back telemetry, a.k.a. downlink and uplink, respectively

presented: one is Load Shift Keying (LSK) for AIMDs, and the other is Frequency Shift Keying (FSK) for RFID applications.

4.2   Design of Power Transfer Links that Need to Support Back Telemetry

As mentioned in Sect. 2.1.5, high-quality factors, Q, are desired for WPT coils to maximize Power Transfer Efficiency (PTE) and Power Delivered to the Load (PDL). However, large Q-factors make it more difficult to transmit data through the same inductive links by limiting the inductive link bandwidth and consequently the achievable data transmission rate.

An isolated series R-L-C circuit has a transient response which decays exponentially at a rate of $e^{-t\frac{R}{2L}}$. It means that at the end of R-L-C excitation, it takes $\frac{2L}{R} = 2Q\frac{1}{\omega} = \frac{Q}{\pi}\frac{1}{f}$ seconds, or $\frac{Q}{\pi}$ cycles, until the oscillation amplitude decays to 37% of its initial value.

Although the bit decay time in a WPT link has a more complex expression which depends on the number of coils used and the coupling coefficient between them [17], this approximation is useful to predict that the larger the coil Q-factor, the longer it takes to transfer each bit. The bit decay time limits the data rate as the data transmitter should wait until the last bit diminishes below a certain threshold before sending a new one. When one bit interferes with the subsequent bit, the received signal is distorted, making it more difficult to distinguish 1's from 0's, rendering the communication less reliable. This is an undesired phenomenon known as Intersymbol Interference (ISI).

Therefore, achieving high PTE and PDL goes against achieving high data rate transmission, and there needs to be a compromise in the link design to achieve the design targets. Consequently, the data rate that is achieved through WPT links is often considerably lower than the dedicated RF data links. In applications such as RFID, which do not need considerable power transfer, the coil's Q-factors are intentionally lowered, thus limiting the efficiency to achieve the desired data rate, which is equally important, if not more (see Sect. 4.3.2).

4.3 Examples of Implementation

4.3.1 Load Shift Keying (LSK)

Load Shift Keying (LSK) is one of the most popular passive modulation schemes in which digital information is transmitted through discrete changes of the load impedance. This technique is used in RFID, wireless sensors, and biomedical applications [10]. The key to its success lies in its very low-power consumption, simple implementation, and robustness when there is sufficient coupling between the coils.

This modulation is represented in Fig. 4.2. In the Rx, the input impedance of the Rx-circuit, Z_{MN}, is changed in accordance with the digital data. These changes are seen from the Tx through the reflected impedance which was calculated in Sect. 2.1.3:

$$(2.6) \Rightarrow Z_{Rx\text{-}Tx_{ref}} = \frac{\omega^2 M_{Tx\text{-}Rx}^2}{R_{Rx} + j\omega L_{Rx} + Z_{MN}} = \frac{\omega^2 k_{Tx\text{-}Rx}^2 L_{Tx} L_{Rx}}{R_{Rx} + j\omega L_{Rx} + Z_{MN}}.$$

The larger the variations in Z_{MN}, the easier to detect the changes in $Z_{Rx\text{-}Tx_{ref}}$ and the more robust the communication is. For the same reason, this modulation scheme fails in links with low coupling coefficient because the changes in Z_{MN} tend to become imperceptible on the Tx side in the presence of various sources of noise and interference.

The Z_{MN} modulation can be divided into resistor switching and capacitor switching. In the resistor switching LSK, only the real part of Z_{MN} is altered while maintaining the Rx at resonance. In the capacitor switching LSK, the imaginary part

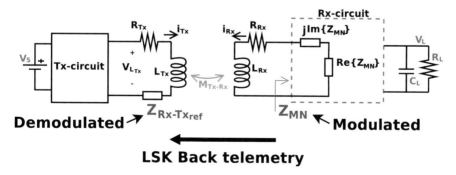

Fig. 4.2 Back telemetry using Load Shift Keying (LSK)

of Z_{MN} is altered, thus detuning the Rx. This passive modulation is carried out while the power is being transmitted, thus affecting the power transmission to a greater or lesser extent. For instance, a change in $Im\{Z_{MN}\}$ causes a great change in the reflected impedance, which is desirable for communication but strongly deteriorates the PTE.

Both capacitive and resistive load modulation can be implemented in varied ways. Some examples of the state-of-the-art are presented in Fig. 4.3 [3–16]. In all the configurations presented in Fig. 4.3, the switch S_m is in charge of modulating Z_{MN}. In Fig. 4.3a, b, the input capacitance of the rectifier is altered (capacitor switching), thus detuning the Rx. All the other examples in Fig. 4.3 are resistor switching since only the real part of Z_{MN} is intentionally changed. However, due to changes in the connection of parasitic capacitance, even in resistor switching, the resonance could be undesirably affected if it is not taken into account in the design process. In Fig. 4.3c, d, the input resistance of the rectifier is changed by connecting a resistor in parallel. In most practical cases of Fig. 4.3c, d, $R_m = 0$, i.e., the input of the rectifier is short-circuited. In Fig. 4.3e, the rectifier is disconnected from the Rx coil; therefore the rectifier input resistance approaches open circuit, instead of shorting as in Fig. 4.3c, d. In Fig. 4.3f, g, the rectifier load resistance is changed, instead of modifying its input. Any change in the rectifier load also affects its input impedance as will be further analyzed in Chap. 6. Finally, in Fig. 4.3h, i, the rectifier architecture is modified with back telemetry data, e.g., from a full-wave rectifier to a voltage doubler, which also affects Z_{MN}. In Fig. 4.3h a passive rectifier is used, while in Fig. 4.3i the diodes were substituted by MOS switches in a circuit known as an active rectifier. The influence of changing the rectifier architecture in the Rx-circuit input impedance, Z_{MN}, is further addressed in Sect. 6.3.2.

In the examples of Fig. 4.3, only two distinct symbols are used for binary data transmission (S_m ON for "1" or OFF for "0"). Additionally, the symbol, i.e., the reflected impedance, that is obtained when the load is connected to the Rx coil, depends on the load power consumption, e.g., when S_m is ON in Fig. 4.3g. To increase the data rate, to have symbols that are more different from each other, and/or to have more than one symbol that is independent of the circuit loading, a multilevel LSK can be implemented by introducing extra symbols. Two examples of multilevel LSK are presented in Fig. 4.4. In Fig. 4.4a, the typical approach of short-circuiting the rectifier input, Fig. 4.3c, is combined with a modification in the rectifier load (Fig. 4.3g). In this scheme three symbols can be generated: rectifier mode (S_{m1} OFF and S_{m2} ON), shorted coil (S_{m1} ON and S_{m2} does not care), and open load (S_{m1} OFF and S_{m2} OFF). In Fig. 4.4b the approach introduced in Fig. 4.3c is used but in this case with two different resistances, R_{m1} and R_{m2}. Additionally, in the particular implementation presented in [18], a dedicated coil in the Rx is used for back telemetry in order to decouple it from the power transmission (and forward telemetry), as shown in Fig. 4.4b. In that case, there are two reflected impedances in the Tx, one from L_{Rx} and another from L_{LSK}, and only the latter is modulated.

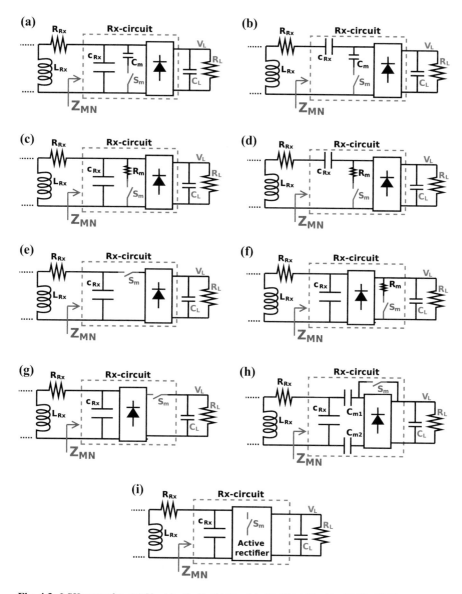

Fig. 4.3 LSK examples. (**a**) Used in [3, 4]. (**b**) Used in [5, 6] (**c**) Used in [7–9]. (**d**) Used in [14]. (**e**) Used in [10]. (**f**) Used in [11]. (**g**) Used in [12]. (**h**) Used in [15, 16]. (**i**) Used in [13]

Even though multilevel LSK can increase the data rate, it inevitably uses symbols that are closer to one another, increasing the probability of error bits. Therefore, using these techniques is only recommended when the noise and interference are low and there is sufficient coupling between the Tx and Rx coils.

4.3.1.1 Example of Use in AIMDs

In AIMDs, LSK is the most popular scheme for passive back telemetry. In this
section, we present the particular implementation proposed in [19] of the multilevel
approach that was introduced in Fig. 4.4a.

The block diagram of the WPT system intended for AIMDs is presented in
Fig. 4.5.

A class E power amplifier drives the Tx resonant tank. The forward telemetry
can be implemented by modulating the amplitude, frequency, or phase of the power

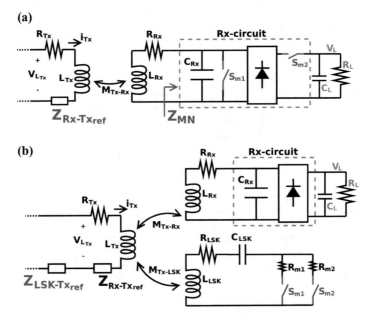

Fig. 4.4 Multilevel LSK examples. (**a**) Used/proposed in [19] (**b**) Used/proposed in [18]

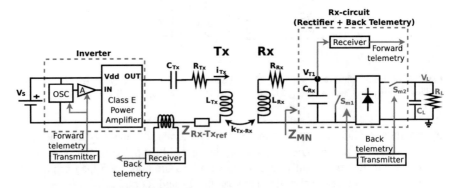

Fig. 4.5 Example of back telemetry using LSK in AIMDs

Table 4.1 Specifications of the inductive link used for multilevel LSK back telemetry

Parameter	Tx	Rx
Litz wire diameter (mm)	0.25	0.15
Number of strands	40	7
Distance between Tx and Rx (mm)	$D_{Tx\text{-}Rx} = 10$	
Coupling between Tx and Rx	$k_{Tx\text{-}Rx} \simeq 0.1$	
Radius (internal/external) (mm)	7/22	3.5/11
Carrier frequency	$f = 500$ kHz	
ESR at the carrier frequency (Ω)	$R_{Tx} = 2.45$	$R_{Rx} = 7.22$
Self-inductance (μH)	$L_{Tx} = 45$	$L_{Rx} = 13$
Resonance capacitor (nF)	$C_{Tx} = 2.25$	$C_{Rx} = 7.8$

Table 4.2 $Z_{Rx\text{-}Tx_{ref}}$ calculation for the example presented in Fig. 4.5 and Table 4.1

STATE		Z_{MN}		$Z_{Rx\text{-}Tx_{ref}}$	$|Z_{Rx\text{-}Tx_{ref}}|$
S_{m1}	S_{m2}	$Re\{Z_{MN}\}$	$Im\{Z_{MN}\}$	(2.6)	Modulus
OFF	ON	Depends on R_L	$\frac{-j}{\omega C_{Rx}} = \omega L_{Rx}$	Depends on R_L	Depends on R_L
OFF	OFF	0	$\frac{-j}{\omega C_{Rx}} = \omega L_{Rx}$	7.99 Ω	7.99 Ω
ON	Does not care	0	0	$(0.24 - 1.37j)$ Ω	1.39 Ω

carrier with the stream of data and detecting these changes on the Rx side. Back telemetry is implemented by the previously introduced multilevel LSK scheme in Fig. 4.4a.

The inductive link used in this example is specified in Table 4.1.

The values of $Z_{Rx\text{-}Tx_{ref}}$ for this example are calculated in Table 4.2. As can be seen, two different symbols that are independent of the load power consumption, R_L, can be generated with $|Z_{Rx\text{-}Tx_{ref}}|_0 = 1.39\ \Omega$ and $|Z_{Rx\text{-}Tx_{ref}}|_1 = 7.99\ \Omega$.

The Rx-circuit for this prototype was fabricated in the AMI-0.5 μm standard n-well CMOS process. Measurement results are presented in Fig. 4.6, for both switches, S_{m1} and S_{m2}, indicating the switch state, the voltage V_{T1} in one terminal of L_{Rx}, the output voltage, V_L, and the Tx current, i_{Tx}. The latter is the one sensed in the Tx to decode back telemetry.

4.3.2 Frequency Shift Keying (FSK)

Frequency Shift Keying (FSK) is a digital Frequency Modulation (FM) scheme in which the information is transmitted through discrete frequency changes in an RF carrier signal. Therefore, unlike LSK, FSK relies on the continuous transmission of a carrier signal, thus having a larger power consumption. Furthermore, traditional FSK requires power-hungry blocks such as Phase-Locked Loop (PLL) and

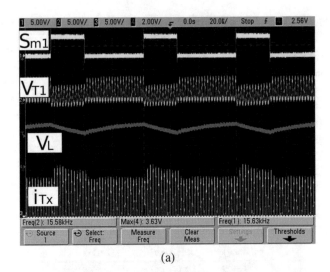

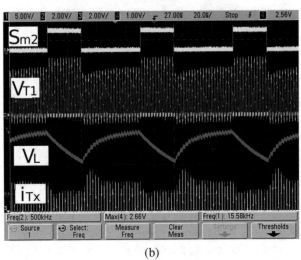

Fig. 4.6 Measurement result of the WPT system with back telemetry presented in Fig. 4.5 and Table 4.1. ($^{©}$2008 IEEE Reprinted, with permission, from [19]). (**a**) Effect of S_{m1} on the output voltage, V_L, and the Tx current, i_{Tx}. (**b**) Effect of S_{m2} on the output voltage, V_L, and the Tx current, i_{Tx}

oscillators, increasing the overall power consumption and complexity. Thus, FSK is appropriate for forward telemetry, as the Tx, which has to modulate the carrier, is less constrained in size and power [20]. However, FSK is also used for back telemetry in systems with low coupling coefficient, such as RFID applications, where the coupling coefficient is too weak to detect changes in the reflected impedance, preventing us from using LSK [21].

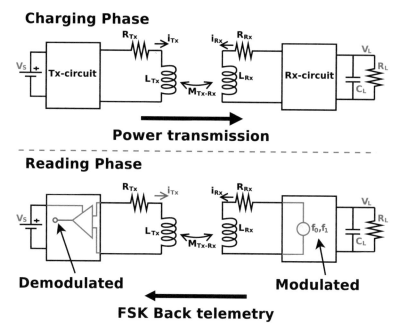

Fig. 4.7 Back telemetry using FSK. Widely used in RFID applications

When FSK is used in back telemetry, the power transmission is usually paused to avoid interference between the two. Therefore, the system is transmitting power from the Tx to the Rx during certain periods and transmitting data from the Rx to the Tx in other periods using the stored energy in the Rx. The FSK scheme used for back telemetry is represented in Fig. 4.7.

In this approach, assuming that the coupling coefficient is such that there is no frequency-splitting effect (studied in Sect. 2.1.7), at least one symbol is transmitted out of resonance as each one is represented by a different frequency. This results in each symbol generating signals with different amplitudes on the Tx side, which is shown in the following example.

4.3.2.1   Example of Using FSK in Low-Frequency RFID

Automatic identification procedures are nowadays used in a wide variety of applications, providing information about people, animals, and goods [22]. Radio Frequency Identification (RFID) is a contactless identification system which, unlike the barcode labels, can be reprogrammed and has larger storage capacity. For that reason, RFID is currently used in many applications such as access control, transportation payments (bus, trains or to collect tolls on highways), and animal identification, and its market is a fast-growing sector [22].

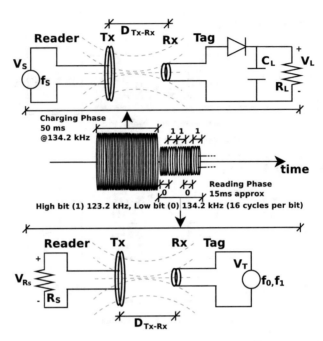

Fig. 4.8 Scheme of a half-duplex RFID system with FSK response. (©2016 IEEE Reprinted, with permission, from [24])

The animal identification is one of the oldest uses of RFID, and it is used both for pet and livestock identification. In this section, we present an example of a system used for cattle identification.

This application uses passive tags (function without a battery) and operates in half-duplex mode with FSK modulation at relatively low-frequency range (134.2 kHz power carrier, 134.2 and 123.2 kHz response frequencies for 0's and 1's, respectively) [23]. The system operation and equivalent schemes for the charging and reading phases are graphically depicted in Fig. 4.8. As can be seen, it consists of a 50 ms charging phase when C_L is charged followed by the reading phase of approximately 15 ms, when the tag responds via FSK while powered from the energy stored in C_L.

In this system, the maximum read distance can be limited either by the charging or the reading phase. If the voltage available in the storage capacitor (C_L in Fig. 4.8) after the charging phase (50 ms) is not enough for the tag to work, it will not respond. The response will also fail if the voltage in the storage capacitor, during the response time, falls below a minimum value required for data transmission by the tag. Actual values for a particular example are provided next. Additionally, if the reader is too far from the tag, the received response may be too weak and thus incorrectly decoded due to reader noise.

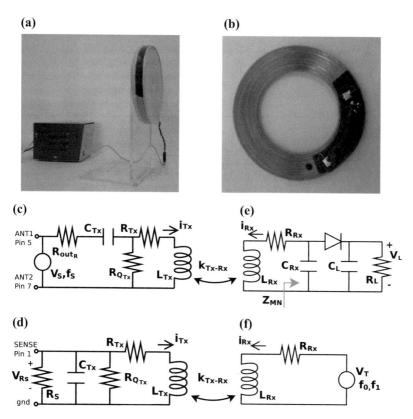

Fig. 4.9 System model. The $R_{Q_{Tx}}$ is added to limit the Tx coil quality factor, Q_{Tx}. The half-wave rectifier (diode and C_L) and R_L model the tag load, while R_S models the reader input impedance during the reading phase. V_S is the reader's driver signal and V_T the tag response signal. R_{out} is the reader's driver output resistance. The pins indicated in **c**, **d** correspond to the TMS3705. (**a**) Reader (TMS3705). ($^{©}$2016 IEEE Reprinted, with permission, from [24].) (**b**) Commercial tag (RI-INL-R9QM). ($^{©}$2016 IEEE Reprinted, with permission, from [24]) (**c**) Reader model during the charging phase. (**d**) Reader model during the reading phase. (**e**) Tag model during the charging phase. (**f**) Tag model during the reading phase

The reader used in this example is shown in Fig. 4.9a. It is based on a TMS3705 transponder base station IC from Texas Instruments (TI). The schematic diagrams are presented in Fig. 4.9c, d for the charging and reading phases, respectively.

The tag is the RI-INL-R9QM from TI and it is shown in Fig. 4.9b. The schematic diagrams showed in Fig. 4.9e, f were derived from the tag datasheet and the guidelines of the International Committee for Animal Recording (ICAR) [25].

Table 4.3 shows the component values for the model of Fig. 4.9, which were measured and estimated for the presented example. The values of R_{Tx}, R_{Rx}, L_{Tx}, L_{Rx}, and R_{out_R} were measured. The $R_{Q_{Tx}}$ was added to reduce the Tx quality factor, from $\sim$23 (without $R_{Q_{Tx}}$) to $\sim$9 (with $R_{Q_{Tx}}$) avoiding ISI, as discussed in Sect. 4.2. The reader generates a 5 Vp (peak voltage) square wave; thus the 6.37 Vp

Table 4.3 Component values of Fig. 4.9

Reader (Tx)	Tag (Rx)
Carrier frequency $f = 134.2\,\text{kHz}$	Low bit frequency $f_0 = 134.2\,\text{kHz}$
	High bit frequency $f_1 = 123.2\,\text{kHz}$
Charging phase duration 50 ms	Reading phase duration $\sim 15\,\text{ms}$
	(16 cycles per bit)
$R_{Tx} = 15.9\,\Omega$ @ 134.2 kHz	$R_{Rx} = 44.59\,\Omega$ @ 134.2 kHz
$R_{Tx} = 15.1\,\Omega$ @ 123.2 kHz	$R_{Rx} = 42.75\,\Omega$ @ 123.2 kHz
$R_{Q_{Tx}} = 5\,\text{k}\Omega$	
$L_{Tx} = 443\,\mu\text{H}$	$L_{Rx} = 2.49\,\text{mH}$
$Q_{Tx} \simeq \dfrac{\omega L_{Tx}}{R_{Tx} + \frac{(\omega L_{Tx})^2}{R_{Q_{Tx}}}} = 8.6$ @ 134.2 kHz	$Q_{Rx} = 47.1$ @ 134.2 kHz
$= 9.0$ @ 123.2 kHz	$Q_{Rx} = 45.1$ @ 123.2 kHz
C_{Tx} and C_{Rx} to resonate at 134.2 kHz	
$R_S = 47\,\text{k}\Omega$	$R_L = 2.2\,\text{M}\Omega$
	$C_L = 68\,\text{nF}$
$V_S = 6.37\,\text{V}_p$	$V_T = 9\,\text{V}_p$
$R_{out_R} = 11.5\,\Omega$	
Radius 6 cm	Radius 1.35 cm

fundamental of this square wave is considered, since higher-order harmonics will be filtered by the system. The peak voltage of the tag's driver (V_T in Table 4.3) was experimentally determined so that the tag equivalent circuit generates the same field strength as an actual tag. The other values were estimated from ICAR [25] guidelines and component datasheets.

In this model, the tag charging is considered successful if V_L is higher than 5 V at the end of the charging phase (based on ICAR guidelines [25]). The amplitude received by the reader from the tag (V_{R_S} Fig. 4.9d) should be high enough in order not to be affected by noise and to be correctly demodulated. It was estimated that the Rx is able to decode the data, i.e., detect the signal frequency during the reading phase, if the received peak amplitude, V_{R_S}, is greater than $10\,\text{mV}_p$.

Using this model, the voltage V_L at the end of the reading phase was calculated, and it is plotted in Fig. 4.10a. The amplitude of V_{R_S}, for the 1's and 0's, was also calculated and is presented in Fig. 4.10b.

Based on the model, both 1's and 0's amplitudes are high enough ($> 10\,\text{mV}$) to be decoded for distances lower than 25 cm (Fig. 4.10b). It can be seen that the bit that operates out of resonance (high bit) is the one with the lower amplitude. Regarding V_L, based on the model, the tag receives enough power if the distance is lower than 20 cm ($V_L > 5\,\text{V}$). Therefore, here it is the charging phase that limits the tag read distance.

In the actual system, the tag is correctly read at distances up to 16 cm, which is slightly less than the distance predicted by the model. In order to prove that it is the charging phase which is limiting the read distance, the Rx (tag) response signal strength was measured, as presented in Fig. 4.11, for $D_{Tx\text{-}Rx} = 16\,\text{cm}$ (tag

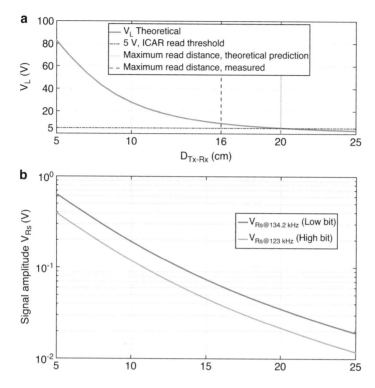

Fig. 4.10 (**a**) Calculated output voltage, V_L (Fig. 4.9e), at the end of the charging phase. (**b**) Calculated sensed voltage, V_{R_S} (Fig. 4.9d), during reading phase

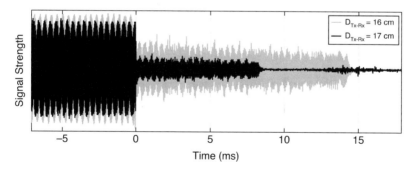

Fig. 4.11 Rx (tag) signal strength when it is at 16 and 17 cm distance from the Tx (reader). (©2016 IEEE Reprinted, with permission, from [24])

successfully read) and $D_{Tx\text{-}Rx} = 17\,\text{cm}$ (tag not read). In Fig. 4.11, we can see the end of the charging phase followed by the tag response. At $D_{Tx\text{-}Rx} = 16\,\text{cm}$, the energy available in the tag was enough to complete the tag ID transmission ($\sim15\,\text{ms}$). However, at $D_{Tx\text{-}Rx} = 17\,\text{cm}$, the tag was unable to complete the transmission, and for longer distances, no response was observed at all. This suggests that the read distance is limited by the charging phase as predicted by the model.

Therefore, in order to increase the read range, the charging phase should be improved. In [24], an additional resonator was placed between the reader and the tag to improve the power transmission (charging phase) to convert the original 2-coil WPT link, discussed in Sect. 2.1, to a 3-coil link, discussed in Sect. 2.2.1. This extra coil inevitably affects the reading phase as well. By using that extra coil, the read distance was improved from 16 to 43 cm (2.7 times improvement in reading distance), proving that using additional resonators can improve certain telemetry systems. This example will be further addressed in Sect. 8.1.

References

1. R.R. Harrison, P.T. Watkins, R.J. Kier, R.O. Lovejoy, D.J. Black, B. Greger, F. Solzbacher, A low-power integrated circuit for a wireless 100-electrode neural recording system. IEEE J. Solid-State Circuits **42**(1), 123–133 (2006)
2. U.-M. Jow, M. Ghovanloo, Optimization of data coils in a multiband wireless link for neuroprosthetic implantable devices. IEEE Trans. Biomed. Circuits Syst. **4**(5), 301–310 (2010)
3. M. Najjarzadegan, I. Ghotbi, S.J. Ashtiani, O. Shoaei, A double-carrier wireless power and data telemetry for implantable biomedical systems, in *IEEE International Symposium on Circuits and Systems* (IEEE, 2016), pp. 2038–2041
4. Y. Lin, K. Tang, An inductive power and data telemetry subsystem with fast transient low dropout regulator for biomedical implants. IEEE Trans. Biomed. Circuits Syst. **10**(2), 435–444 (2016). ISSN 1940-9990. https://doi.org/10.1109/TBCAS.2015.2447526
5. S. Mao, H. Wang, C. Zhu, Z.-H. Mao, M. Sun, Simultaneous wireless power transfer and data communication using synchronous pulse-controlled load modulation. Measurement **109**, 316–325 (2017)
6. D. Huwig, P. Wambsganß, Digitally controlled synchronous bridge-rectifier for wireless power receivers, in *Annual IEEE Conference on Applied Power Electronics Conference and Exposition* (IEEE, 2013), pp. 2598–2603
7. S. Mandal, R. Sarpeshkar, Power-efficient impedance-modulation wireless data links for biomedical implants. IEEE Trans. Biomed. Circuits Syst. **2**(4), 301–315 (2008)
8. H. Lee, An auto-reconfigurable 2 × /4× AC-DC regulator for wirelessly powered biomedical implants with 28% link efficiency enhancement. IEEE Trans. VLSI Syst. **24**(4), 1598–1602 (2016). ISSN 1063-8210. https://doi.org/10.1109/TVLSI.2015.2452918
9. H.-M. Lee, M. Ghovanloo, An integrated power-efficient active rectifier with offset-controlled high speed comparators for inductively powered applications. IEEE Trans. Circuits Syst. I **58**(8), 1749–1760 (2011)
10. A. Trigui, S. Hached, A.C. Ammari, Y. Savaria, M. Sawan, Maximizing data transmission rate for implantable devices over a single inductive link: Methodological review. IEEE Rev. Biomed. Eng. **12**, 72–87 (2018)
11. N. Chaimanonart, M.A. Suster, W.H. Ko, D.J. Young, Two-channel data telemetry with remote RF powering for high-performance wireless MEMS strain sensing applications, in *IEEE Sensors* (IEEE, 2005), p. 4

12. G. Wang, W. Liu, M. Sivaprakasam, G.A. Kendir, Design and analysis of an adaptive transcutaneous power telemetry for biomedical implants. IEEE Trans. Circuits Syst. I **52**(10), 2109–2117 (2005). ISSN 1549-8328. https://doi.org/10.1109/TCSI.2005.852923

13. X. Li, C.Y. Tsui, W.H. Ki, A 13.56 MHz wireless power transfer system with reconfigurable resonant regulating rectifier and wireless power control for implantable medical devices. IEEE J. Solid-State Circuits **50**(4), 978–989 (2015)

14. D. Ahn, S. Kim, J. Moon, I.-K. Cho, Wireless power transfer with automatic feedback control of load resistance transformation. IEEE Trans. Power Electron. **31**(11), 7876–7886 (2015)

15. Y. Hu, M. Sawan, A fully integrated low-power BPSK demodulator for implantable medical devices. IEEE Trans. Circuits Syst. I **52**(12), 2552–2562 (2005)

16. Z. Tang, B. Smith, J.H. Schild, P.H. Peckham, Data transmission from an implantable biotelemeter by load-shift keying using circuit configuration modulator. IEEE Trans. Biomed. Eng. **42**(5), 524–528 (1995)

17. A.P. French, *Vibrations and Waves* (CRC Press, 2017)

18. W. Xu, Z. Luo, S. Sonkusale, Fully digital BPSK demodulator and multilevel LSK back telemetry for biomedical implant transceivers. IEEE Trans. Circuits Syst. II **56**(9), 714–718 (2009)

19. M. Ghovanloo, S. Atluri, An integrated full-wave CMOS rectifier with built-in back telemetry for RFID and implantable biomedical applications. IEEE Trans. Circuits Syst. I **55**(10), 3328–3334 (2008)

20. M. M. Ahmadi, S. Ghandi, A class-E power amplifier with wideband FSK modulation for inductive power and data transmission to medical implants. IEEE Sensors J. **18**(17), 7242–7252 (2018)

21. C. Gong, D. Liu, Z. Miao, W. Wang, M. Li, An NFC on two-coil WPT link for implantable biomedical sensors under ultra-weak coupling. Sensors **17**(6), 1358 (2017)

22. K. Finkenzeller, *RFID Handbook: Fundamentals and Applications in Contactless Smart Cards, Radio Frequency Identification and Near-Field Communication* (Wiley, 2010)

23. ISO, ISO 11785 radio frequency identification of animals – technical concept. Standard, International Organization for Standardization (1996)

24. P. Pérez-Nicoli, A. Rodríguez-Esteva, F. Silveira, Bidirectional analysis and design of RFID using an additional resonant coil to enhance read range. IEEE Trans. Microw. Theory Techn. **64**(7), 2357–2367 (2016). ISSN 0018-9480. https://doi.org/10.1109/TMTT.2016.2573275

25. ICAR, International agreement of recording practices (2008) www.icar.org

Chapter 5
Achieving the Optimum Operating Point (OOP)

5.1 Introduction

To achieve high Power Transfer Efficiency (PTE) of the link, η_{Link}, and high power delivered to the Rx-circuit, P_{MN}, it is important to match the load impedance of the Rx coil, Z_{MN}, with the rest of the inductive link (see Fig. 5.1). The higher the η_{Link}, the higher the total PTE, $\eta_{TOT} = \eta_{Tx} \times \eta_{Link} \times \eta_{Rx}$, and the larger the P_{MN}, the larger the Power Delivered to the Load (PDL), $P_L = P_{MN} \times \eta_{Rx}$. As discussed in Sect. 2.1.6, a considerably higher η_{Link} and P_{MN} are obtained if the Rx coil is resonating, i.e., the input impedance of the Rx-circuit, Z_{MN}, cancels the Rx coil reactance. Additionally, the real part of Z_{MN} can also be adapted to maximize η_{Link} or P_{MN}, thus maximizing the PTE or the PDL of the system. In this chapter, we address the influence of the input impedance of the Rx-circuit, Z_{MN}, in the PTE and in the PDL.

Let us start with the example presented in Table 5.1 (Fig. 5.2) for which η_{Link} and P_{MN} are plotted as functions of the real and imaginary part of Z_{MN}. Regarding the imaginary part, $Im\{Z_{MN}\}$, both, η_{Link} and P_{MN}, are maximized when the $Im\{Z_{MN}\} = -1000 = -\omega L_{Rx}$ which is the Rx resonance condition. This is an expected result as the advantages of operating at resonance for η_{Link} and P_{MN} were addressed in Sect. 2.1.6.

As for the real part, $Re\{Z_{MN}\}$, as can be foreseen from Figs. 5.3 and 5.4, and will be demonstrated in this chapter, two optimum values for $Re\{Z_{MN}\}$ exist. One of these values maximizes η_{Link}, achieving the Maximum Efficiency Point (MEP), while the other maximizes P_{MN}, achieving the Maximum Power Point (MPP). The Optimum Operating Point (OOP) is used to refer to both, MEP and MPP, depending on the desired link optimization paradigm, which is in turn linked to the application.

The Z_{MN} can be adjusted using, for instance, a matching network, to achieve the MEP or MPP. As exemplified by Figs. 5.3 and 5.4, being far from the MEP or MPP can involve large losses of PTE or PDL, respectively. This highlights the importance

© The Author(s), under exclusive license to Springer Nature Switzerland AG 2021
P. Pérez-Nicoli et al., *Inductive Links for Wireless Power Transfer*,
https://doi.org/10.1007/978-3-030-65477-1_5

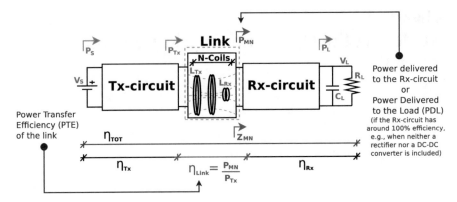

Fig. 5.1 Figure to recall the definition of Z_{MN}, P_{MN} and η_{Link}

of optimizing the Z_{MN} and achieving the desired OOP, which are studied in this chapter.

This chapter is organized as follows: First, in Sects. 5.2 and 5.3, the MEP and MPP are addressed in a resonant 2-coil link, respectively. A comparison between the MEP and MPP is then presented in Sect. 5.4. The MEP and MPP in a 3-coil link are addressed in Sect. 5.5. Then, after presenting the optimum values for $Re\{Z_{MN}\}$, the use of matching network to achieve the desired Z_{MN} is introduced in Sect. 5.6. Finally, 2-coil and 3-coil links are compared in Sect. 5.7 followed by an exemplar design of a 3-coil link in Sect. 5.8.

5.2 Maximum Efficiency Point (MEP) in 2-Coil Links

The inductive link efficiency, η_{Link}, for a resonant 2-coil link was deduced in Sect. 2.1.3:

$$\eta_{Link} = \frac{Q_{Rx\text{-}L}}{Q_L} \frac{k_{Tx\text{-}Rx}^2 Q_{Tx} Q_{Rx\text{-}L}}{k_{Tx\text{-}Rx}^2 Q_{Tx} Q_{Rx\text{-}L} + 1}, \tag{2.13}$$

where $Q_{Rx\text{-}L}$ is

$$Q_{Rx\text{-}L} = \frac{Q_{Rx} Q_L}{Q_{Rx} + Q_L}. \tag{2.11}$$

Rewriting (2.13) explicitly as a function of Q_{Rx}, η_{Link} results in the following expression:

$$\left.\begin{array}{c}(2.13)\\(2.11)\end{array}\right\} \Rightarrow \eta_{Link} = \left(\frac{Q_{Rx}}{Q_{Rx} + Q_L}\right) \frac{k_{Tx\text{-}Rx}^2 Q_{Tx} Q_{Rx} Q_L}{k_{Tx\text{-}Rx}^2 Q_{Tx} Q_{Rx} Q_L + Q_{Rx} + Q_L} \tag{5.1}$$

Table 5.1   Introductory example. The calculation script is available in the supplementary material, file Sec510

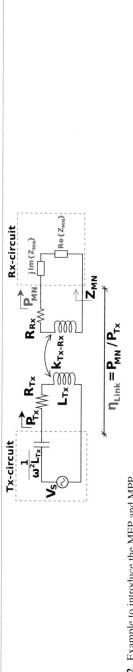

Fig. 5.2   Example to introduce the MEP and MPP

Parameters for the example:

$R_{Tx} = R_{Rx} = 5\,\Omega$; $Q_{Tx} = Q_{Rx} = 200$; $k_{Tx\text{-}Rx} = 0.01$

$f = 13.56\,\text{MHz}$; $V_S = 1\,\text{V}$

MEP   **MPP**

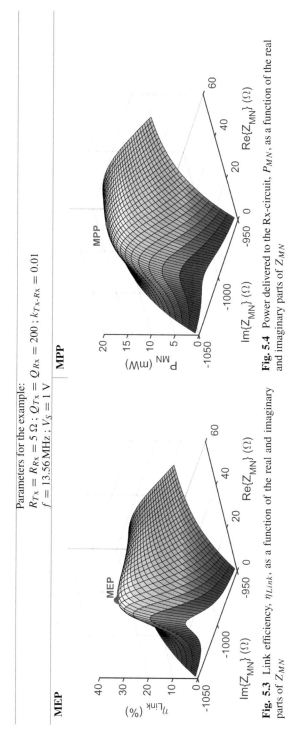

Fig. 5.3   Link efficiency, η_{Link}, as a function of the real and imaginary parts of Z_{MN}

Fig. 5.4   Power delivered to the Rx-circuit, P_{MN}, as a function of the real and imaginary parts of Z_{MN}

The Q_L that maximizes η_{Link}, i.e., achieves the MEP, can be found by calculating the partial derivative of η_{Link} with respect to Q_L as presented in (5.2). A step-by-step deduction of (5.2) can be found in Appendix B.1:

$$\frac{\partial \eta_{Link}}{\partial Q_L} = \frac{k_{Tx\text{-}Rx}^2 Q_{Tx} Q_{Rx}^3 [Q_{Rx}^2 - Q_L^2 (k_{Tx\text{-}Rx}^2 Q_{Tx} Q_{Rx} + 1)]}{(Q_{Rx} + Q_L)^2 (k_{Tx\text{-}Rx}^2 Q_{Tx} Q_{Rx} Q_L + Q_{Rx} + Q_L)^2} \Rightarrow$$

$$\frac{\partial \eta_{Link}}{\partial Q_L} = 0 \iff Q_L = \frac{Q_{Rx}}{\sqrt{k_{Tx\text{-}Rx}^2 Q_{Tx} Q_{Rx} + 1}} = Q_{Lopt\text{-}\eta} \tag{5.2}$$

Therefore the MEP is achieved when $Q_L = Q_{Lopt\text{-}\eta}$:

$$Q_{Lopt\text{-}\eta} = \frac{Q_{Rx}}{\sqrt{k_{Tx\text{-}Rx}^2 Q_{Tx} Q_{Rx} + 1}}. \tag{5.3}$$

Using the definition of $Q_{Rx} = \omega L_{Rx}/R_{Rx}$ and

$$Q_L = \frac{\omega L_{Rx}}{Re\{Z_{MN}\}}, \tag{2.4}$$

the MEP can be written as a function of $Re\{Z_{MN}\}$ as:

$$Re\{Z_{MN_{opt\text{-}\eta}}\} = R_{Rx}\sqrt{k_{Tx\text{-}Rx}^2 Q_{Tx} Q_{Rx} + 1}. \tag{5.4}$$

Evaluating η_{Link} at $Q_{Lopt\text{-}\eta}$, the maximum value of η_{Link} is found:

$$\eta_{Link\,max} = \frac{k_{Tx\text{-}Rx}^2 Q_{Tx} Q_{Rx}}{(\sqrt{k_{Tx\text{-}Rx}^2 Q_{Tx} Q_{Rx} + 1} + 1)^2}. \tag{5.5}$$

The deduction of (5.5) can be found in Appendix B.2.

To understand the existence of this optimum value, let us recall how η_{Link} was calculated in Sect. 2.1.3. The η_{Link} was obtained calculating the efficiency of the Rx coil, $\eta_{L_{Rx}}$ (2.2), and the Tx coil, $\eta_{L_{Tx}}$ (2.12), $\eta_{Link} = \eta_{L_{Rx}} \times \eta_{L_{Tx}}$. Thus, the efficiency calculation can be recalled from Figs. 2.3, (2.2), and (2.12):

$$\eta_{L_{Rx}} = \frac{Re\{Z_{MN}\}}{Re\{Z_{MN}\} + R_{Rx}} \tag{2.2}$$

$$\eta_{L_{Tx}} = \frac{R_{Rx\text{-}Tx_{ref}}}{R_{Rx\text{-}Tx_{ref}} + R_{Tx}}. \tag{2.12}$$

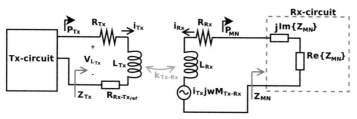

Fig. 2.3 The 2-coil link model showing reflected resistance on the Tx side and induced voltage on the Rx side (repeated from page 21)

where the reflected resistance, $R_{Rx\text{-}Tx_{ref}}$, is:

$$R_{Rx\text{-}Tx_{ref}} = \frac{\omega^2 k_{Tx\text{-}Rx}^2 L_{Tx} L_{Rx}}{R_{Rx} + Re\{Z_{MN}\}} \tag{2.8}$$

The $\eta_{L_{Rx}}$ (2.2) was obtained as the ratio between the power delivered to the Rx-circuit, $Re\{Z_{MN}\}i_{Rx}^2/2$, and the total power received, $(Re\{Z_{MN}\} + R_{Rx})i_{Rx}^2/2$, where i_{Rx} is the peak current through the Rx coil. Analogously, the $\eta_{L_{Tx}}$ was calculated as the ratio between the power transferred to the Rx coil, $R_{Rx\text{-}Tx_{ref}}i_{Tx}^2/2$, and the total power received by the Tx coil, $(R_{Rx\text{-}Tx_{ref}} + R_{Tx})i_{Tx}^2/2$, where i_{Tx} is the peak current through the Tx coil. From (2.2), (2.12), (2.8), and the definition of quality factors, the η_{Link} (2.13) was obtained (see Sect. 2.1.3).

The existence of the optimum $Re\{Z_{MN}\}$ can be intuitively understood in two ways: The first is to try to analyze in the model of Fig. 2.3 how a change in $Re\{Z_{MN}\}$ would affect $\eta_{L_{Rx}}$ and $\eta_{L_{Tx}}$. If $Re\{Z_{MN}\}$ increases, the loss in the series R_{Rx} resistor will be smaller, and thus the efficiency $\eta_{L_{Rx}}$ would be higher (a higher proportion of the power dissipated in the Rx side would be dissipated in $Re\{Z_{MN}\}$). Note that we are considering the PTE and not the amount of Power Delivered to the Load (PDL), which will be addressed in Sect. 5.3 and leads to a different optimum condition. On the other hand, if we consider the Tx side, the value of $Re\{Z_{MN}\}$ would affect its performance through the reflected resistance value $R_{Rx\text{-}Tx_{ref}}$. When $Re\{Z_{MN}\}$ increases, the reflected resistance $R_{Rx\text{-}Tx_{ref}}$ decreases, as shown in (2.8), and an effect that is opposite to the one in the Rx side happens. As $Re\{Z_{MN}\}$ increases and $R_{Rx\text{-}Tx_{ref}}$ decreases, the loss in R_{Tx} becomes proportionally more significative with respect to the power delivered to $R_{Rx\text{-}Tx_{ref}}$ (which is the power delivered to the Rx), and the efficiency $\eta_{L_{Tx}}$ decreases. Therefore, changing $Re\{Z_{MN}\}$ has opposite effects on $\eta_{L_{Tx}}$ and $\eta_{L_{Rx}}$, which leads to the existence of an optimum, midway, value that provides the best trade-off between an increase in $\eta_{L_{Rx}}$ and a decrease in $\eta_{L_{Tx}}$, thus maximizing the product $\eta_{L_{Tx}} \times \eta_{L_{Rx}}$ and the total link efficiency.

A second, more rigorous, way to understand the dependences just presented on $Re\{Z_{MN}\}$, which leads to the existence of the optimum, is to realize that the $\eta_{L_{Rx}}$ expression of (2.2) is a monotonically increasing function of $Re\{Z_{MN}\}$, while from (2.12) and (2.8), it can be seen that $\eta_{L_{Tx}}$ is a monotonically decreasing function of $Re\{Z_{MN}\}$. Therefore, an optimum value for $Re\{Z_{MN}\}$ exists, which maximizes $\eta_{Link} = \eta_{L_{Tx}} \times \eta_{L_{Rx}}$.

To complete this section, the example of Fig. 5.5 is presented, which is a series-series resonant link, whose parameters are specified in the caption. The $\eta_{L_{Tx}}$, $\eta_{L_{Rx}}$ and η_{Link} for the example of Fig. 5.5 are presented in Fig. 5.6. In this example, with a series resonant Rx coil, $Re\{Z_{MN}\} = R_L$. The objective of this example is to illustrate, numerically, the previous analysis. The results showed in Fig. 5.6 agree with the analysis presented in this section as $\eta_{L_{Rx}}$ is a monotonously increasing function of $Re\{Z_{MN}\}$, $\eta_{L_{Tx}}$ is a monotonously decreasing function of $Re\{Z_{MN}\}$, and an optimum R_L exists which is calculated in (5.6).

$$(5.4) \Rightarrow Re\{Z_{MN_{opt-\eta}}\} = R_{Rx}\sqrt{k_{Tx\text{-}Rx}{}^2 Q_{Tx} Q_{Rx} + 1}$$

$$= 5\,\Omega\sqrt{0.01^2 \times 200 \times 200 + 1} \tag{5.6}$$

$$\simeq 11.18\,\Omega$$

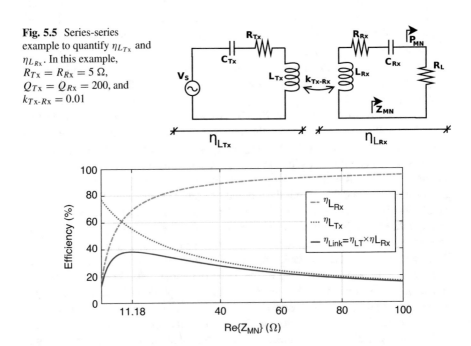

Fig. 5.5 Series-series example to quantify $\eta_{L_{Tx}}$ and $\eta_{L_{Rx}}$. In this example, $R_{Tx} = R_{Rx} = 5\,\Omega$, $Q_{Tx} = Q_{Rx} = 200$, and $k_{Tx\text{-}Rx} = 0.01$

Fig. 5.6 Quantification of $\eta_{L_{Tx}}$, $\eta_{L_{Rx}}$ and $\eta_{Link} = \eta_{L_{Tx}} \times \eta_{L_{Rx}}$, for the example presented in Fig. 5.5. The $Re\{Z_{MN_{opt-\eta}}\}$ for this case is calculated in (5.6). The calculation script is available in the supplementary material, file Sec520

5.3   Maximum Power Point (MPP) in 2-Coil Links

In the previous section, the value of $Re\{Z_{MN}\}$ that maximizes η_{Link} was calculated. In this section, we deduce and analyze the existence of a different value of $Re\{Z_{MN}\}$ that maximizes the power delivered to the Rx-circuit, P_{MN}, which is yet another important parameter in the design of an inductive wireless link, as explained in Chap. 2.

As was summarized in Table 2.3, contrary to the η_{Link}, the expression for P_{MN} depends on the Tx-circuit used. Hence, the $Re\{Z_{MN}\}$ that maximizes P_{MN} also depends on the Tx-circuit. First, in Sect. 5.3.1, the case with a voltage source and series resonant capacitor in the Tx-circuit is addressed. Then, in Sect. 5.3.2, the case with a current source and series resonant capacitor is considered.

5.3.1   MPP, Tx-circuit with a Voltage Source and a Series Resonant Capacitor

For the case of a voltage source with a series resonant Tx coil (Fig. 2.4), the P_{MN} is the one presented in (2.17), which was deduced in Appendix A.1.

$$P_{MN} = \frac{V_S^2}{2R_{Tx}} \frac{Q_{Rx\text{-}L}}{Q_L} \frac{k_{Tx\text{-}Rx}{}^2 Q_{Tx} Q_{Rx\text{-}L}}{(k_{Tx\text{-}Rx}{}^2 Q_{Tx} Q_{Rx\text{-}L} + 1)^2} \qquad (2.17)$$

Rewriting P_{MN} as:

$$\left.\begin{array}{r} (2.17) \\ Q_{Rx\text{-}L} = \frac{Q_{Rx} Q_L}{Q_{Rx} + Q_L} \qquad (2.11) \end{array}\right\} \Rightarrow$$

$$\qquad\qquad (5.7)$$

$$P_{MN} = \frac{V_S^2}{2R_{Tx}} \frac{k_{Tx\text{-}Rx}{}^2 Q_{Tx} Q_{Rx}{}^2 Q_L}{(k_{Tx\text{-}Rx}{}^2 Q_{Tx} Q_{Rx} Q_L + Q_{Rx} + Q_L)^2},$$

Fig. 2.4 Voltage source and series resonant Tx (repeated from page 24)

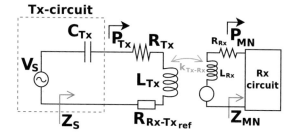

the Q_L that maximizes P_{MN} can be found as:

$$\frac{\partial P_{MN}}{\partial Q_L} = 0 \Longleftrightarrow Q_L = \frac{Q_{Rx}}{k_{Tx\text{-}Rx}^2 Q_{Tx} Q_{Rx} + 1} = Q_{Lopt\text{-}PMN}. \tag{5.8}$$

A detailed deduction of (5.7) and (5.8) can be found in Appendix B.3.

Finally, the $Re\{Z_{MN_{opt\text{-}PMN}}\}$ that achieves the MPP is:

$$\left.\begin{array}{c} Q_{Lopt\text{-}PMN} = \dfrac{Q_{Rx}}{k_{Tx\text{-}Rx}^2 Q_{Tx} Q_{Rx} + 1} \\[2mm] Q_{Rx} = \dfrac{\omega L_{Rx}}{R_{Rx}} \\[2mm] Q_L = \dfrac{\omega L_{Rx}}{Re\{Z_{MN}\}} \end{array}\right\} \Rightarrow \tag{5.9}$$

$$Re\{Z_{MN_{opt\text{-}PMN}}\} = R_{Rx}(k_{Tx\text{-}Rx}^2 Q_{Tx} Q_{Rx} + 1).$$

Evaluating P_{MN} (5.7) at $Q_{Lopt\text{-}PMN}$ (5.8), the maximum P_{MN} is obtained:

$$P_{MN\max} = \frac{V_S^2}{8 R_{Tx}} \frac{k_{Tx\text{-}Rx}^2 Q_{Tx} Q_{Rx}}{(1 + k_{Tx\text{-}Rx}^2 Q_{Tx} Q_{Rx})}. \tag{5.10}$$

The deduction of (5.10) can be found in Appendix B.4.

The $Re\{Z_{MN_{opt\text{-}PMN}}\}$ can also be deduced calculating the Thévenin equivalent model of the link and using the maximum power transfer theorem [1]. The Thévenin equivalent model of the 2-coil link with a voltage source and a resonant series Tx coil is presented in Fig. 5.7.

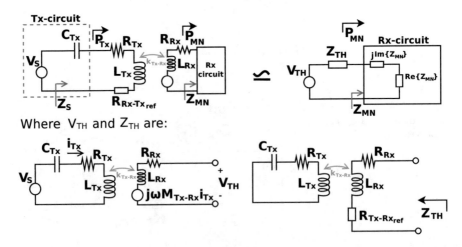

Fig. 5.7 Thévenin equivalent model of 2-coil link with voltage source and series resonant Tx

In Fig. 5.7, the Thévenin-equivalent voltage, V_{TH}, is the open-circuit voltage at the output terminal of the inductive link:

$$\left. \begin{array}{l} V_{TH} = j\omega M_{Tx\text{-}Rx} i_{Tx} \\ i_{Tx} = \dfrac{V_S}{R_{Tx}} \end{array} \right\} \Rightarrow V_{TH} = j\omega M_{Tx\text{-}Rx} \dfrac{V_S}{R_{Tx}}. \tag{5.11}$$

The equivalent impedance, Z_{TH}, shown in Fig. 5.7, is the impedance seen looking into the output terminal of the inductive link when the voltage source, V_S, is replaced with a short circuit. Z_{TH} can be calculated reusing the reflected resistance expression of (2.8), but in this case from the Rx to the Tx. This deduction is presented in (5.12):

$$\begin{aligned} Z_{TH} &= R_{Rx} + R_{Tx\text{-}Rx_{ref}} + j\omega L_{Rx} \\ &= R_{Rx} + \frac{\omega^2 k_{Tx\text{-}Rx}^2 L_{Tx} L_{Rx}}{R_{Tx}} + j\omega L_{Rx} \\ &= R_{Rx}(1 + k_{Tx\text{-}Rx}^2 Q_{Tx} Q_{Rx}) + j\omega L_{Rx} \end{aligned} \tag{5.12}$$

Using the maximum power transfer theorem, P_{MN} is maximum when Z_{MN} is equal to the conjugate of Z_{TH}:

$$Z_{MN_{opt\text{-}P_{MN}}} = \overline{Z_{TH}}$$

$$Re\{Z_{MN_{opt\text{-}P_{MN}}}\} + j Im\{Z_{MN_{opt\text{-}P_{MN}}}\} = R_{Rx}(1 + k_{Tx\text{-}Rx}^2 Q_{Tx} Q_{Rx}) - j\omega L_{Rx} \tag{5.13}$$

Therefore, MPP is achieved when the Rx is resonating, $j Im\{Z_{MN}\} = -j\omega L_{Rx}$, and $Re\{Z_{MN}\} = R_{Rx}(1 + k_{Tx\text{-}Rx}^2 Q_{Tx} Q_{Rx})$, which agrees with (5.9).

The P_{MN} at the $Re\{Z_{MN_{opt\text{-}P_{MN}}}\}$, from the Thévenin equivalent circuit, is:

$$\begin{aligned} P_{MN\,\max} &= \frac{|V_{TH}/2|^2}{2Re\{Z_{TH}\}} = \frac{\left(\omega M_{Tx\text{-}Rx} \dfrac{V_S}{R_{Tx}}/2\right)^2}{2R_{Rx}(1 + k_{Tx\text{-}Rx}^2 Q_{Tx} Q_{Rx})} \\ &= \frac{V_S^2}{8R_{Tx}} \frac{k_{Tx\text{-}Rx}^2 \omega L_{Tx} \omega L_{Rx}}{R_{Rx} R_{Tx}(1 + k_{Tx\text{-}Rx}^2 Q_{Tx} Q_{Rx})} \\ &= \frac{V_S^2}{8R_{Tx}} \frac{k_{Tx\text{-}Rx}^2 Q_{Tx} Q_{Rx}}{(1 + k_{Tx\text{-}Rx}^2 Q_{Tx} Q_{Rx})} \end{aligned} \tag{5.14}$$

which agrees with (5.10).

The analysis of this section, presented for a voltage source Tx with a series resonant capacitor (Fig. 2.4), is also applicable for its dual, i.e., the case of a current source Tx with a parallel resonant capacitor (Fig. 2.6), due to the similarity in the expression of P_{MN} in both cases (Table 2.3) in what regards the dependence with $Q_{Rx\text{-}L}$ and Q_L, which are the terms that depend on $Re\{Z_{MN}\}$.

5.3.2 MPP, Tx-circuit with a Current Source and a Series Resonant Capacitor

Analogously to Section 5.3.1, in this section, the Q_L that maximizes P_{MN} is calculated, here for the current source Tx with a series resonant capacitor, shown in Fig. 2.6.

The $Q_{Lopt\text{-}P_{MN}}$ for this case is presented in (5.15); its step-by-step deduction can be found in Appendix B.5:

$$\left.\begin{array}{l} P_{MN} = \frac{I_S^2}{2} R_{Tx} \frac{Q_{Rx\text{-}L}}{Q_L} k_{Tx\text{-}Rx}^2 Q_{Tx} Q_{Rx\text{-}L} \quad (2.19) \\[6pt] Q_{Rx\text{-}L} = \frac{Q_{Rx}Q_L}{Q_{Rx}+Q_L} \quad (2.11) \end{array}\right\} \Rightarrow$$

$$P_{MN} = \frac{I_S^2}{2} R_{Tx} \frac{Q_{Rx}}{Q_{Rx} + Q_L} k_{Tx\text{-}Rx}^2 Q_{Tx} \frac{Q_{Rx}Q_L}{Q_{Rx} + Q_L} \Rightarrow \tag{5.15}$$

$$\frac{\partial P_{MN}}{\partial Q_L} = \frac{I_S^2}{2} R_{Tx} k_{Tx\text{-}Rx}^2 Q_{Tx} Q_{Rx}^2 \frac{Q_{Rx}^2 - Q_L^2}{(Q_{Rx} + Q_L)^4}$$

$$\frac{\partial P_{MN}}{\partial Q_L} = 0 \iff Q_L = Q_{Rx} = Q_{Lopt\text{-}P_{MN}}$$

From (5.15), the $Re\{Z_{MN_{opt\text{-}P_{MN}}}\}$ can be deduced as:

$$\left.\begin{array}{l} Q_{Lopt\text{-}P_{MN}} = Q_{Rx} \\[6pt] Q_{Rx} = \frac{\omega L_{Rx}}{R_{Rx}} \\[6pt] Q_L = \frac{\omega L_{Rx}}{Re\{Z_{MN}\}} \end{array}\right\} \Rightarrow Re\{Z_{MN_{opt\text{-}P_{MN}}}\} = R_{Rx}. \tag{5.16}$$

It should be noted that, as mentioned above, the $Re\{Z_{MN_{opt\text{-}P_{MN}}}\}$ that maximizes P_{MN} depends on the Tx-circuit used. With a voltage source, in the previous

Fig. 2.6 Current source and series resonant Tx. (repeated from page 24)

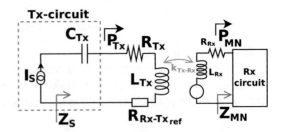

section, the $Re\{Z_{MN_{opt-P_{MN}}}\}$ of (5.9) was deduced, which is different from the result obtained in this Sect. (5.16).

The maximum P_{MN}, at $Re\{Z_{MN_{opt-P_{MN}}}\}$, is:

$$P_{MN\,max} = \frac{I_S^2}{8} R_{Tx} k_{Tx\text{-}Rx}^2 Q_{Tx} Q_{Rx},\qquad(5.17)$$

the step-by-step deduction of (5.17) is presented in Appendix B.6.

Both (5.16) and (5.17) could also be deduced using the Thévenin equivalent model of Fig. 5.8.

Applying, as in Sect. 5.3.1, the theorem of maximum power transfer, $Re\{Z_{MN_{opt-P_{MN}}}\}$ is equal to $Re\{Z_{TH}\}$:

$$Re\{Z_{MN_{opt-P_{MN}}}\} = Re\{Z_{TH}\} = R_{Rx} + \underbrace{R_{Tx\text{-}Rx_{ref}}}_{=0} = R_{Rx},\qquad(5.18)$$

which agrees with (5.16).

Using the Thévenin model, $P_{MN\,max}$ can be calculated as:

$$\begin{aligned}P_{MN\,max} &= \frac{|V_{TH}/2|^2}{2Re\{Z_{TH}\}} = \frac{(\omega M_{Tx\text{-}Rx} I_S/2)^2}{2R_{Rx}}\\[2mm] &= \frac{I_S^2}{8}\frac{k_{Tx\text{-}Rx}^2 \omega L_{Tx}\omega L_{Rx}}{R_{Rx}}\frac{R_{Tx}}{R_{Tx}}\\[2mm] &= \frac{I_S^2}{8} R_{Tx} k_{Tx\text{-}Rx}^2 Q_{Tx} Q_{Rx},\end{aligned}\qquad(5.19)$$

which agrees with (5.17).

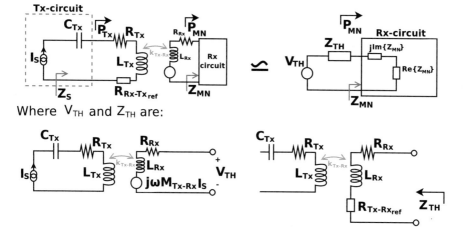

Fig. 5.8 Thévenin equivalent model of 2-coil link with current source and series resonant Tx

The analysis of this section, presented for a current source Tx with a series resonant capacitor (Fig. 2.5), is also applicable for its dual, i.e., a voltage source Tx with a parallel resonant capacitor, Fig. 2.5, due to the similarity in the expression of P_{MN} in both cases (Table 2.3) in what regards the dependence with $Q_{Rx\text{-}L}$ and Q_L, which are the terms that depend on $Re\{Z_{MN}\}$.

5.4 Choosing Between MEP and MPP

It was proven, in Sects. 5.2 and 5.3, that two different optimum values for $Re\{Z_{MN}\}$ exist, one of them maximizes η_{Link}, while the other maximizes P_{MN}. The value of $Re\{Z_{MN}\}$ can be adjusted modifying the Rx-circuit, as will be discussed in Sect. 5.6. In this section, the MEP and the MPP are compared indicating how to choose between them.

Let us start with the example presented in Table 5.2, which has a voltage source Tx with a series resonant capacitor (Fig. 5.9). The coil parameters selected for this example are presented in Table 5.2, and a desired P_{MN} of 500 mW is assumed for wirelessly charging a wearable device, such as a smartwatch. In this example, if $Re\{Z_{MN}\}$ is adjusted to operate at the MEP, the system requires a $V_S = 7.8$ V to achieve the desired $P_{MN} = 500$ mW. This is an Optimum Operating Point with maximum efficiency, $\eta_{Link} = 81.9\%$. However, to operate at this point, the Tx-circuit must be able to generate the required voltage of 7.8 V, which may not be possible depending on the Tx-circuit used. If the Tx-circuit operates at a maximum output voltage of, for example, 5 V from a USB port, the desired $P_{MN} = 500$ mW cannot be achieved by operating at the MEP. If this is the case, it would be necessary to operate at the MPP instead since the transferred power is not at a level for temperature rise to be a major concern here. In the same example presented in Table 5.2, when $Re\{Z_{MN}\}$ is adjusted to operate at the MPP, a $V_S = 4.5$ V is required to achieve the desired $P_{MN} = 500$ mW. In this MPP, the desired P_{MN} is achieved, at the cost of η_{Link} being lower at 49%. In Table 5.2, the main parameters are summarized for the MEP and the MPP.

As can be seen from the example of Table 5.2, at the MEP, the reflected resistance in the Tx, $R_{Tx\text{-}Rx_{ref}}$, is larger than at the MPP. Also at MEP, $R_{Tx\text{-}Rx_{ref}} \gg R_{Tx}$ ($45.25\Omega \gg 5\Omega$), minimizing the loss in the parasitic resistance of L_{Tx} but also requiring a larger V_S to deliver the desired $P_{MN} = 500$ mW to the Rx-circuit. At the MPP, the reflected resistance is much lower, $R_{Tx\text{-}Rx_{ref}} \simeq R_{Tx}$ ($4.9\Omega \simeq 5\Omega$), to maximize the power delivered to the Rx-circuit with lower V_S, although the η_{Link} is reduced by almost the same amount of power being dissipated in R_{Tx}.

The MEP ensures an efficient use of power and operating at this point is important to have low-power dissipation, low-temperature operation, and extend Tx lifetime when it is powered from a battery. These could be the highest priorities in an AIMD application. The MEP would be the first choice, as high efficiency is always desirable. Therefore, the system should operate at the MEP as long as the Tx-circuit is capable of achieving the desired P_{MN} in those conditions. If this

Table 5.2 Comparison between MEP and MPP, with a voltage source Tx and series resonant capacitor. The calculation script is available in the supplementary material, file Sec540

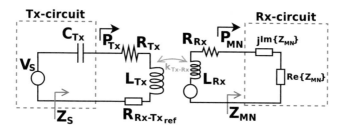

Fig. 5.9 Example to compare MEP and MPP, with a voltage source Tx and series resonant capacitor

	Parameters used in this example:	
	$R_{Tx} = R_{Rx} = 5\,\Omega$; $Q_{Tx} = Q_{Rx} = 200$;	
	$k_{Tx\text{-}Rx} = 0.05$	
	Desired $P_{MN} = 500\,\text{mW}$	
	Resonance is assumed, $\omega L_{Tx} = 1/(\omega C_{Tx})$	
	and $\omega L_{Rx} = -Im\{Z_{MN}\}$	
Parameter	**MEP**	**MPP**
$Re\{Z_{MN}\}$	$(5.4) \Rightarrow 50.25\,\Omega$	$(5.9) \Rightarrow 505\,\Omega$
Q_L	$(5.3) \Rightarrow 19.9$	$(5.8) \Rightarrow 1.98$
η_{Link}	$(2.13) \Rightarrow 81.9\%$	$(2.13) \Rightarrow 49\%$
Required V_S to achieve $P_{MN} = 500\,\text{mW}$	7.8 V	4.5 V
P_{MN} at $V_S = 5$ V	204 mW	619 mW
$R_{Tx\text{-}Rx_{ref}}$	$(2.10) \Rightarrow 45.25\,\Omega$	$(2.10) \Rightarrow 4.9\,\Omega$
$P_{Tx} = \frac{P_{MN}}{\eta_{Link}} = \frac{V_S{}^2}{2(R_{Tx}+R_{Tx\text{-}Rx_{ref}})}$	248.8 mW	1262 mW

is not the case, and the desired P_{MN} is not achieved even with the Tx-circuit at its maximum amplitude, then $Re\{Z_{MN}\}$ should be adjusted to operate closer to the MPP, at the cost of lower efficiency, η_{Link}. Alternatively, a different Tx-circuit should be considered.

A similar example is presented in Table 5.3, for the case of a current source Tx and a series resonant capacitor. In that example, the MPP has a higher reflected impedance, $R_{Tx\text{-}Rx_{ref}}$, which means larger power at a fix I_S. Analogous to the voltage source Tx case, the MEP should be selected as long as the Tx-circuit is able to generate the required current to achieve the desired P_{MN}. If it is not the case, operating closer to the MPP should be considered.

There is a case where MEP and MPP tend to coincide, as discussed next. To achieve this, let us address the existence of these two $Re\{Z_{MN}\}$ values, which maximize efficiency, η_{Link}, and power, P_{MN}. A change in $Re\{Z_{MN}\}$ modifies not only the η_{Link} but also the reflected resistance, $R_{Tx\text{-}Rx_{ref}}$, and thus the Tx-

Table 5.3 Comparison between MEP and MPP, with a current source Tx and series resonant capacitor. The calculation script is available in the supplementary material, file Sec540

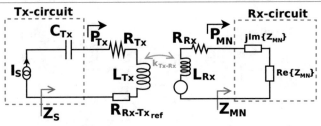

Fig. 5.10 Example to compare MEP and MPP, with a current source Tx and series resonant capacitor

Parameters for the example:		
$R_{Tx} = R_{Rx} = 5\,\Omega$; $Q_{Tx} = Q_{Rx} = 200$;		
$k_{Tx\text{-}Rx} = 0.05$		
Desired $P_{MN} = 500\,\text{mW}$		
Resonance is assumed, $\omega L_{Tx} = 1/(\omega C_{Tx})$		
and $\omega L_{Rx} = -Im\{Z_{MN}\}$		
Parameter	**MEP**	**MPP**
$Re\{Z_{MN}\}$	(5.4) $\Rightarrow$ 50.25 Ω	(5.16) $\Rightarrow$ 5 Ω
Q_L	(5.3) $\Rightarrow$ 19.9	(5.15) $\Rightarrow$ 200
η_{Link}	(2.13) $\Rightarrow$ 81.9%	(2.13) $\Rightarrow$ 49%
Required I_S to achieve $P_{MN} = 500\,\text{mW}$	156 mA	89 mA
P_{MN} at $I_S = 100\,\text{mA}$	205.8 mW	625 mW
$R_{Tx\text{-}Rx_{ref}}$	(2.10) $\Rightarrow$ 45.25 Ω	(2.10) $\Rightarrow$ 250 Ω
At $I_S = 100\,\text{mA}$ $P_{Tx} = \frac{P_{MN}}{\eta_{Link}} = \frac{1}{2}I_S^2(R_{Tx} + R_{Tx\text{-}Rx_{ref}})$	251.2 mW	1275 mW

circuit output power, P_{Tx}. Therefore, the point that maximizes η_{Link} may reflect a Tx-circuit load resistance which limits its output power, i.e., too large load resistance with a voltage source Tx or too low load resistance with a current source Tx (Fig. 5.10). However, when the coils are far apart with low $k_{Tx\text{-}Rx}$, the reflected impedance is too low compared with R_{Tx}. In that situation, any change in $Re\{Z_{MN}\}$ is almost imperceptible in the Tx-circuit load resistance, $R_{Tx} + R_{Tx\text{-}Rx_{ref}}$. This means that in a link with low $k_{Tx\text{-}Rx}$, the Tx-circuit output power is almost independent of $Re\{Z_{MN}\}$. Therefore, under this condition, maximizing the η_{Link} should be the same as maximizing the P_{MN}. This conclusion agrees with the equations deduced for $Re\{Z_{MN_{opt\text{-}\eta}}\}$ (5.4) and $Re\{Z_{MN_{opt\text{-}P_{MN}}}\}$ ((5.9) and (5.16)). When $k_{Tx\text{-}Rx}$ approaches zero, both $Re\{Z_{MN_{opt\text{-}\eta}}\}$ and $Re\{Z_{MN_{opt\text{-}P_{MN}}}\}$ tend to R_{Rx}. Thus, in that situation, maximizing η_{Link} is the same as maximizing P_{MN}, and both are achieved by adjusting the Rx-circuit to have $Re\{Z_{MN}\} = R_{Rx}$. This is summarized in (5.20):

$$\lim_{k_{Tx\text{-}Rx}\to 0} \overbrace{Re\{Z_{MN_{opt\text{-}\eta}}\}}^{(5.4)} = \lim_{k_{Tx\text{-}Rx}\to 0} \overbrace{Re\{Z_{MN_{opt\text{-}P_{MN}}}\}}^{(5.9)\text{ and }(5.16)} = R_{Rx}$$

(5.20)

$\Rightarrow$ in loosely coupled links, $k_{Tx\text{-}Rx} \to 0$, the MEP and the MPP are

at the same operating point.

In practice, as can be seen from the deduced expressions for $Re\{Z_{MN_{opt\text{-}\eta}}\}$ and $Re\{Z_{MN_{opt\text{-}P_{MN}}}\}$, (5.4), (5.9), and (5.16), it can be assumed that $Re\{Z_{MN_{opt\text{-}\eta}}\} \simeq Re\{Z_{MN_{opt\text{-}P_{MN}}}\}$ if $k_{Tx\text{-}Rx}{}^2 Q_{Tx} Q_{Rx} \ll 1$, e.g., $k_{Tx\text{-}Rx}{}^2 Q_{Tx} Q_{Rx} \leq 0.1$.

5.5   MEP and MPP in N-Coil Links

So far, in this chapter, only 2-coil links have been considered. However, it was proven in Sect. 2.2 that the use of additional resonant coils in the link can improve η_{Link} and P_{MN}. Those systems with multiple coils also have an optimum $Re\{Z_{MN_{opt\text{-}\eta}}\}$ that maximizes η_{Link}, achieving the MEP, and another optimum $Re\{Z_{MN_{opt\text{-}P_{MN}}}\}$ that maximizes P_{MN}, achieving the MPP. The analytical expressions for $Re\{Z_{MN_{opt\text{-}\eta}}\}$ and $Re\{Z_{MN_{opt\text{-}P_{MN}}}\}$ depend on the number of coils used in the link, and the more coils are used, the more complex the analytical expressions are.

In this section, by way of example, the $Re\{Z_{MN_{opt\text{-}\eta}}\}$ and $Re\{Z_{MN_{opt\text{-}P_{MN}}}\}$ for a 3-coil link are presented, as well as the $\eta_{Link\,max}$ and $P_{MN\,max}$ for this case.

The η_{Link} for a 3-coil link was deduced in Sect. 2.2.1:

$$\eta_{Link} = \frac{Q_{Rx\text{-}L}}{Q_L} \frac{k_{A\text{-}Rx}{}^2 Q_A Q_{Rx\text{-}L}}{(k_{A\text{-}Rx}{}^2 Q_A Q_{Rx\text{-}L} + 1)} \frac{k_{Tx\text{-}A}{}^2 Q_{Tx} Q_A}{(k_{Tx\text{-}A}{}^2 Q_{Tx} Q_A + k_{A\text{-}Rx}{}^2 Q_A Q_{Rx\text{-}L} + 1)}.$$

(2.32)

The Q_L that maximizes η_{Link} in this case is:

$$\frac{\partial \eta_{Link}}{\partial Q_L} = 0 \Leftrightarrow Q_L = Q_{L opt\text{-}\eta}$$

$$Q_{L opt\text{-}\eta} = Q_{Rx}\sqrt{\frac{k_{Tx\text{-}A}{}^2 Q_{Tx} Q_A + 1}{(k_{A\text{-}Rx}{}^2 Q_A Q_{Rx} + 1)(k_{A\text{-}Rx}{}^2 Q_A Q_{Rx} + k_{Tx\text{-}A}{}^2 Q_{Tx} Q_A + 1)}}$$

(5.21)

A step-by-step deduction of (5.21) can be found in Appendix B.7.

Evaluating η_{Link} at $Q_{L opt\text{-}\eta}$, the $\eta_{Link\,max}$ for this 3-coil case is found:

$$\eta_{Link\,max} = \frac{A^2 - 1}{(A + 1)^2}$$

$$\text{Where,} \quad A = \sqrt{1 + \frac{k_{Tx\text{-}A}{}^2 Q_{Tx} Q_A k_{A\text{-}Rx}{}^2 Q_A Q_{Rx}}{k_{Tx\text{-}A}{}^2 Q_{Tx} Q_A + k_{A\text{-}Rx}{}^2 Q_A Q_{Rx} + 1}} \tag{5.22}$$

A step-by-step deduction of (5.22) is presented in Appendix B.8.

From (5.21) and the quality factor definitions, $Re\{Z_{MN_{opt\text{-}\eta}}\}$ can be deduced as follows:

$$\left. \begin{array}{l} Q_{L\,opt\text{-}\eta} \quad (5.21) \\ Q_{Rx} = \dfrac{\omega L_{Rx}}{R_{Rx}} \\ Q_L = \dfrac{\omega L_{Rx}}{Re\{Z_{MN}\}} \end{array} \right\} \Rightarrow$$

$$Re\{Z_{MN_{opt\text{-}\eta}}\} = R_{Rx} \sqrt{\frac{(k_{A\text{-}Rx}{}^2 Q_A Q_{Rx}+1)(k_{A\text{-}Rx}{}^2 Q_A Q_{Rx}+k_{Tx\text{-}A}{}^2 Q_{Tx} Q_A+1)}{k_{Tx\text{-}A}{}^2 Q_{Tx} Q_A+1}}. \tag{5.23}$$

Regarding the $Re\{Z_{MN_{opt\text{-}P_{MN}}}\}$, similar to the 2-coil case, it depends on the Tx-circuit used. If a voltage source with a series resonant capacitor is used in the Tx-circuit, $Q_{L\,opt\text{-}P_{MN}}$ can be deduced as follows (step-by-step deduction in Appendix B.9):

$$P_{MN} = \frac{V_S{}^2}{2R_{Tx}} \frac{Q_{Rx\text{-}L}}{Q_L} \frac{(k_{A\text{-}Rx}{}^2 Q_A Q_{Rx\text{-}L})(k_{Tx\text{-}A}{}^2 Q_{Tx} Q_A)}{(k_{Tx\text{-}A}{}^2 Q_{Tx} Q_A + k_{A\text{-}Rx}{}^2 Q_A Q_{Rx\text{-}L} + 1)^2} \tag{2.34}$$

$$\frac{\partial P_{MN}}{\partial Q_L} = 0 \Leftrightarrow Q_L = Q_{L\,opt\text{-}P_{MN}} = Q_{Rx} \frac{k_{Tx\text{-}A}{}^2 Q_{Tx} Q_A + 1}{k_{Tx\text{-}A}{}^2 Q_{Tx} Q_A + k_{A\text{-}Rx}{}^2 Q_A Q_{Rx} + 1} \tag{5.24}$$

Therefore, $Re\{Z_{MN_{opt\text{-}P_{MN}}}\}$ is:

$$Re\{Z_{MN_{opt\text{-}P_{MN}}}\} = R_{Rx} \frac{k_{Tx\text{-}A}{}^2 Q_{Tx} Q_A + k_{A\text{-}Rx}{}^2 Q_A Q_{Rx} + 1}{k_{Tx\text{-}A}{}^2 Q_{Tx} Q_A + 1}, \tag{5.25}$$

and $P_{MN\,max}$ is:

$$P_{MN\,max} = \frac{V_S^2}{8R_{Tx}} \frac{k_{Tx\text{-}A}{}^2 Q_{Tx} Q_A . k_{A\text{-}Rx}{}^2 Q_A Q_{Rx}}{(k_{Tx\text{-}A}{}^2 Q_{Tx} Q_A + 1)(k_{A\text{-}Rx}{}^2 Q_A Q_{Rx} + k_{Tx\text{-}A}{}^2 Q_{Tx} Q_A + 1)}. \tag{5.26}$$

The deduction of (5.26) is presented in Appendix B.10.

The expressions presented in this section are going to be used in Sect. 5.7 to compare the 2-coil with the 3-coil link and to design a 3-coil link that operates at MEP in Sect. 5.8.

5.6 Using Matching Networks to Achieve the OOP

A matching network can be used to adjust Z_{MN}, achieving resonance and/or the OOP. Matching networks can be implemented in very different ways. Transmission line impedance transformers are often used at high frequencies (at $\gtrsim$ GHz), where the wavelength is comparable to the circuit size. In near-field inductive applications (often in MHz range), lumped matching circuits are more appropriate. Many different lumped matching networks exist like L-Match, π-Match, and T-Match, among other combinations [2]. In the particular case of L-Match, eight different combinations exist using inductors and capacitors, some examples of which are presented in Table 5.4, including the Z_{MN} of each one. Note that the R_L in Table 5.4 could be the load to be powered or the input resistance of the subsequent block, e.g., approximate input resistance of the rectifier.

It is desired that the $Im\{Z_{MN}\}$ does not depend on R_L to keep the inductive link at resonance regardless of the load power consumption (R_L). Therefore, some assumptions were added in Table 5.4 to ensure that the Rx resonance does not depend on R_L.

Each matching network topology has different characteristics, such as frequency response, component values, efficiency, and $Re\{Z_{MN}\}$ proportional or inversely proportional to R_L. Additionally, they are differently affected by parasitic capacitances in their input (e.g., capacitance of the coil) and their output (rectifier input capacitance). Which architecture can be used mainly depends on R_L and $Z_{MN_{opt}}$ ($Z_{MN_{opt-\eta}}$ or $Z_{MN_{opt-P_{MN}}}$) because not every impedance transformation can be achieved with each matching network. Detailed analysis of matching network properties can be found in classical reference works [2, 5].

The series or parallel capacitor, commonly used to achieve resonance at the Rx, can be considered as a particular case of L-Match where one component has zero or infinite impedance. In that case, that particular matching network has only one design variable that is usually set to achieve resonance, $Im\{Z_{MN}\} = -\omega L_{Rx}$, without purposely adjusting $Re\{Z_{MN}\}$. This series or parallel resonant capacitor and even the series-parallel topology of Fig. 5.13, which was applied in [3], are usually referred to as resonant structures instead of matching networks.

The series and parallel resonators are the most typical resonant structures. The η_{Link} and P_{MN} for both (series and parallel) resonant structures were compared previously in Sect. 2.1.6, using the example of Table 2.6. In Sect. 2.1.6, it was highlighted that the series resonator performs better in high-power applications, while the parallel resonator is preferred in low-power applications. In this section, we arrive at the same conclusion from a different perspective by considering the example presented in Table 5.5.

Table 5.4 Examples of different L-Match matching networks and their corresponding Z_{MN}

 Fig. 5.11 Series	$$Z_{MN} = R_L + \frac{1}{j\omega C_S} \qquad (5.27)$$
 Fig. 5.12 Parallel	$$\left. \begin{aligned} Z_{MN} &= \frac{R_L}{1+j\omega C_P R_L} = \frac{R_L(1-j\omega C_P R_L)}{1+(\omega C_P R_L)^2} \\ \text{assuming } (\omega C_P R_L)^2 &>> 1 \\ \Rightarrow Z_{MN} &= \frac{1}{R_L(\omega C_P)^2} - j\frac{1}{\omega C_P} \end{aligned} \right\} \Rightarrow \qquad (5.28)$$
 Fig. 5.13 Series-parallel [3]	$$Z_{MN} = \frac{R_L}{(R_L\omega C_P)^2+1} - j\left(\frac{1}{\omega C_S} + \frac{R_L^2\omega C_P}{(R_L\omega C_P)^2+1} \right)$$ $$\text{assuming } (\omega C_P R_L)^2 >> 1 \Rightarrow$$ $$Z_{MN} = \frac{1}{R_L(\omega C_P)^2} - j\left(\frac{1}{\omega C_P} + \frac{1}{\omega C_S} \right) \qquad (5.29)$$
 Fig. 5.14 C-L [4]	$$Z_{MN} = \frac{1}{j\omega C_S} + \frac{j\omega L_P R_L}{j\omega L_P + R_L}$$ $$\text{assuming } (\omega L_P/R_L)^2 << 1 \qquad (5.30)$$ $$Z_{MN} = \frac{(\omega L_P)^2}{R_L} + j\left(\omega L_P - \frac{1}{\omega C_S} \right)$$

Except for very high-efficiency links where $k_{Tx\text{-}Rx}{}^2 Q_{Tx} Q_{Rx}$ is much greater than 1, typically, the values of $Re\{Z_{MN_{opt\text{-}\eta}}\}$ and $Re\{Z_{MN_{opt\text{-}P_{MN}}}\}$ are similar (i.e., in the same order of magnitude) to R_{Rx}. Note that in the example presented in Table 5.5, where the maximum link efficiency, $\eta_{Link\max}$, is 38.2%, $Re\{Z_{MN_{opt\text{-}\eta}}\}$ and $Re\{Z_{MN_{opt\text{-}P_{MN}}}\}$ are around two times and five times R_{Rx}, respectively. Therefore, when R_L is of the same order of magnitude as R_{Rx}, thus a low-valued R_L and a high-power link, the series resonance imposes a $Re\{Z_{MN}\} = R_L$, which is near the optimum value. However, when R_L is orders of magnitudes larger than R_{Rx}, if a series resonance is used, the $Re\{Z_{MN}\} = R_L$ would be too far from the optimum point, and performance will be poor. In this situation, where $R_L \gg R_{Rx}$, the parallel resonance achieves a $Re\{Z_{MN}\}$ closer to the OOP. In summary, series or parallel resonators are selected based on which one operates closer to the OOP, $Re\{Z_{MN_{opt\text{-}\eta}}\}$ or $Re\{Z_{MN_{opt\text{-}P_{MN}}}\}$. This discussion agrees with

Table 5.5 Comparison between series and parallel Rx resonators. The calculation script is available in the supplementary material, file Sec560

Parameters for the table:

$R_{Tx} = R_{Rx} = 5\,\Omega\,;\, Q_{Tx} = Q_{Rx} = 200\,;\, k_{Tx\text{-}Rx} = 0.01$

| Traditional series-series WPT link already presented in Fig. 2.2a | Traditional series-parallel WPT link already presented in Fig. 2.2b |

Optimum conditions, valid for both, series and parallel Rx resonators:

$$(5.4) \Rightarrow Re\{Z_{MN_{opt\text{-}\eta}}\} = R_{Rx}\sqrt{k_{Tx\text{-}Rx}^2 Q_{Tx}Q_{Rx} + 1} = 11.18\,\Omega$$

$$(5.5) \Rightarrow \eta_{Link\,max} = \frac{k_{Tx\text{-}Rx}^2 Q_{Tx}Q_{Rx}}{(\sqrt{k_{Tx\text{-}Rx}^2 Q_{Tx}Q_{Rx}+1}+1)^2} = 38.2\,\%$$

$$(5.9) \Rightarrow Re\{Z_{MN_{opt\text{-}P_{MN}}}\} = R_{Rx}(k_{Tx\text{-}Rx}^2 Q_{Tx}Q_{Rx} + 1) = 25\,\Omega$$

$$(5.10) \Rightarrow P_{MN\,max} = \frac{V_S^2}{8R_{Tx}}\frac{k_{Tx\text{-}Rx}^2 Q_{Tx}Q_{Rx}}{(1+k_{Tx\text{-}Rx}^2 Q_{Tx}Q_{Rx})} = \begin{matrix} 5\,\text{mW} \to @V_S = 500\,\text{mV} \\ 50\,\text{W} \to @V_S = 50\,\text{V} \end{matrix}$$

SERIES	PARALLEL
	$(5.28) \Rightarrow Re\{Z_{MN}\}=\frac{1}{R_L(\omega C_{Rx})^2}$
	resonance $\Rightarrow \frac{1}{\omega C_{Rx}}=\omega L_{Rx}$ $\Rightarrow$
$Re\{Z_{MN}\} = R_L$	
	$Re\{Z_{MN}\} = Q_{Rx}^2\dfrac{R_{Rx}^2}{R_L}$

Example, $R_L = 200\,\Omega$ and $V_S = 50\,\text{V}$ (High power example)	
SERIES	PARALLEL
$Re\{Z_{MN}\} = 200\,\Omega$	$Re\{Z_{MN}\} = 5\,\text{k}\Omega$
$\eta_{Link} = 8.7\,\%$	$\eta_{Link} = 0.4\,\%$
$P_{MN} = 19.8\,\text{W}$	$P_{MN} = 1\,\text{W}$
Example, $R_L = 3\,\text{k}\Omega$ and $V_S = 500\,\text{mV}$ (Low power example)	
SERIES	PARALLEL
$Re\{Z_{MN}\} = 3\,\text{k}\Omega$	$Re\{Z_{MN}\} = 333\,\Omega$
$\eta_{Link} = 0.7\%$	$\eta_{Link} = 5.5\%$
$P_{MN} = 0.16\,\text{mW}$	$P_{MN} = 1.3\,\text{mW}$

the examples presented in Table 5.5. In the high-power example, where R_L is close to R_{Rx} (less than two orders of magnitude), the series resonator achieves higher η_{Link} and P_{MN} than the parallel resonator. However, in the low-power example, where R_L is much larger, the parallel resonator performs better than the series one.

In the examples presented in Table 5.5, it should be noted that although large improvement can be achieved by correctly selecting the type of resonance circuit, between series and parallel, in all the situations, the η_{Link} and P_{MN} obtained are far from the maximum values of $\eta_{Link\,max}$ and $P_{MN\,max}$. This suggests that using another matching network, e.g., the ones presented in Figs. 5.13 and 5.14 in Table 5.4, the $Re\{Z_{MN}\}$ could be set to the optimum value, and a larger η_{Link}

and/or P_{MN} could be achieved. The matching network should be correctly selected, for instance, based on (5.29), the matching network presented in Fig. 5.13 can only be used if $Re\{Z_{MN_{opt}}\} < R_L$, as $Re\{Z_{MN}\} = \frac{R_L}{(R_L\omega C_P)^2+1}$ is always lower than R_L.

The efficiency of a matching network depends on the quality factor of the components used, $Q_{L_{lumped}} = \omega L / R_{PL}$ for an inductor and $Q_{C_{lumped}} = 1/(\omega C R_{PC})$ for a capacitor (where R_{PL} and R_{PC} are the parasitic series resistance of the inductor and capacitor, respectively) [5]. The higher the component quality factor is, the better the resulting matching network efficiency will be. Matching networks that only use capacitors like the ones shown in Figs. 5.11, 5.12, and 5.13 are preferred as $Q_{C_{lumped}} \gg Q_{L_{lumped}}$ for most practical cases [5, 6].

5.7 Comparing 2-Coil and 3-Coil Links at the MEP

In Sect. 2.3, 2-, 3-, and 4-coil links were compared, showing the advantages of adding resonators in between the main two inductors of wireless links. In that section, the same $Re\{Z_{MN}\}$ (Q_L) was considered in all the cases. However, in this chapter, it was shown that the $Re\{Z_{MN}\}$, which maximizes η_{Link} in a 2-coil link (5.4), is different from the one that maximizes η_{Link} in a 3-coil link (5.23). The same occurs with the P_{MN}, i.e., the expression for $Re\{Z_{MN}\}_{opt-P_{MN}}$ is different in 2-coil links (5.9) and 3-coil links (5.25). When an extra resonator is included in the link, the optimum $Re\{Z_{MN}\}$, which achieves the OOP, changes. Therefore, the comparison presented in Sect. 2.3, where the same $Re\{Z_{MN}\}$ (Q_L) was used in all the links, is made more accurate here.

A more accurate comparison would consider the fact that each link can have its own optimum $Re\{Z_{MN}\}$, set by a specifically designed matching network. We can compare, for instance, the $\eta_{Link\,max}$ deduced in this chapter, which is the link efficiency assuming an optimum $Re\{Z_{MN_{opt-\eta}}\}$.

The $\eta_{Link\,max}$ for 2- and 3- coil links can be written as presented in (5.31); both maximum efficiencies have been already deduced in (5.5) and (5.22), respectively:

$$\eta_{Link\,max} = \frac{A_i^2 - 1}{(A_i + 1)^2},\qquad(5.31)$$

where A_i is:

in 2-coil links $A_{(i=2)} = A_2 = \sqrt{1 + k_{Tx\text{-}Rx}^2 Q_{Tx} Q_{Rx}}$ (5.32)

in 3-coil links $A_{(i=3)} = A_3 = \sqrt{1 + \dfrac{k_{Tx\text{-}A}^2 Q_{Tx} Q_A k_{A\text{-}Rx}^2 Q_A Q_{Rx}}{k_{Tx\text{-}A}^2 Q_{Tx} Q_A + k_{A\text{-}Rx}^2 Q_A Q_{Rx} + 1}}.$

From (5.31), it can be seen that $\eta_{Link\,max}$ is a monotonously increasing function of A_i. Thus when designing the coils for this link, A_2 in 2-coil links and A_3 in 3-

coil links should be maximized. Moreover, adding a resonant coil to increase the link efficiency is beneficial if and only if:

$$\underbrace{k_{Tx\text{-}Rx}^2 Q_{Tx} Q_{Rx}}_{\alpha_2} < \underbrace{\frac{k_{Tx\text{-}A}^2 Q_{Tx} Q_A . k_{A\text{-}Rx}^2 Q_A Q_{Rx}}{k_{Tx\text{-}A}^2 Q_{Tx} Q_A + k_{A\text{-}Rx}^2 Q_A Q_{Rx} + 1}}_{\alpha_3} . \tag{5.33}$$

This inequality provides a very simple method to determine if an additional coil is beneficial, assuming that the MEP is going to be set because the maximum efficiency is desired.

Let us conclude the analysis of this section presenting the numerical example of Table 5.6. The 2-coil link, operating at the MEP, is used as the benchmark to determine if an additional resonator in this link is beneficial or not. The 2-coil link achieves a link efficiency of $\eta_{Link\,max} = 14.5\%$, and the parameter α_2, defined in (5.33), is 0.79. For the 3-coil link, α_3, $Q_{L\,opt\text{-}\eta}$, and η_{Link} are presented as a function of the additional resonator position, for three different additional resonator quality factors, Q_A. In this example, if the $Q_A = 5$ (very low), α_3 is lower than α_2 regardless of the additional resonator position. Therefore, as the inequality of (5.33) is not satisfied, the additional resonator is not beneficial, and it reduces the link efficiency. With a higher Q_A of 50, α_3 becomes higher than α_2 only for some positions of the additional resonator, between 13 and 47 cm. Hence, within that range, the 3-coil link exceeds the efficiency of the 2-coil link. Finally, with a greater Q_A of 200, α_3 is larger than α_2 for any additional resonator position; thus, the inclusion of this third coil is beneficial regardless of its position.

5.8 Design of a 3-Coil Link to Operate at the MEP

The efficiency of a 3-coil link (2.32) depends on the position of the additional resonator, which affects its coupling coefficients with the other two coils, $k_{Tx\text{-}A}$ and $k_{A\text{-}Rx}$:

$$\eta_{Link} = \frac{Q_{Rx\text{-}L}}{Q_L} \frac{k_{A\text{-}Rx}^2 Q_A Q_{Rx\text{-}L}}{(k_{A\text{-}Rx}^2 Q_A Q_{Rx\text{-}L} + 1)} \frac{k_{Tx\text{-}A}^2 Q_{Tx} Q_A}{(k_{Tx\text{-}A}^2 Q_{Tx} Q_A + k_{A\text{-}Rx}^2 Q_A Q_{Rx\text{-}L} + 1)} . \tag{2.32}$$

Assuming three given coils, with a fixed distance between Tx and Rx, $D_{Tx\text{-}Rx}$, an optimum position for the additional resonator, exists, which maximizes η_{Link}. That optimum position depends on $Q_L = \omega L_{Rx}/Re\{Z_{MN}\}$. Let us quantify this on the example presented in Table 5.7, Fig. 5.15. In that example, the η_{Link} is plotted as a function of the Tx to additional resonator distance, for two different Q_L. As can be seen in Fig. 5.16, for each Q_L, an optimum position exists.

Table 5.6 Comparison between 2-coil and 3-coil links operating at the MEP. The calculation script is available in the supplementary material, file Sec570

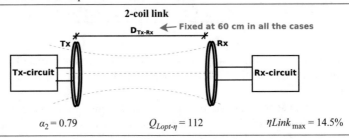

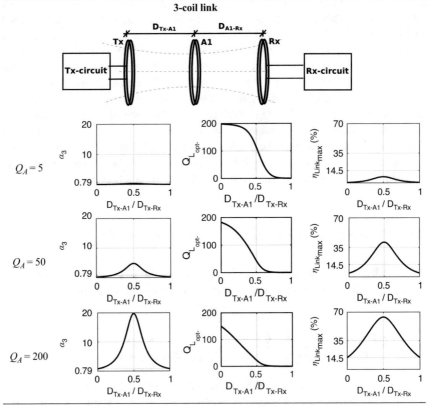

On the other hand, in this chapter, it was proved that an optimum Q_L (5.21) exists which maximizes η_{Link}. This $Q_{Lopt-\eta}$ depends on k_{Tx-A} and k_{A-Rx}; thus it depends on the position of the additional resonator:

Table 5.7 Example of 3-coil link to analyze the design of the matching network. The calculation script is available in the supplementary material, file Sec580

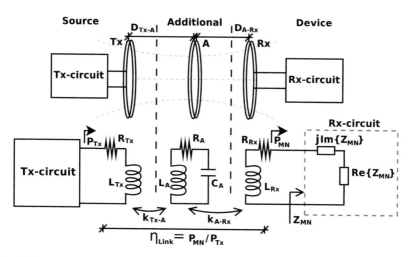

Fig. 5.15 A simplified diagram of a 3-coil WPT link and its lumped circuit model, repeated from Fig. 2.20

Parameters for the example:	
Coils	All identical, with: $Q_{Tx} = Q_A = Q_{Rx} = 200$, Circular shape, diameter 20 cm
Distance	$D_{Tx\text{-}Rx} = 60$ cm, aligned
Coupling coefficients	Are calculated using the approximation (3.16) (Chap. 3)
Rx-circuit	Resonance is assumed, $\omega L_{Rx} = -Im\{Z_{MN}\}$

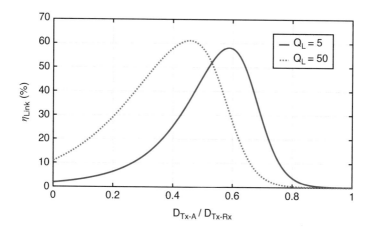

Fig. 5.16 The η_{Link} calculated from (2.32) as a function of the Tx to additional resonator distance, $D_{Tx\text{-}A}$

$$Q_{Lopt\text{-}\eta} = Q_{Rx}\sqrt{\frac{k_{Tx\text{-}A}^2 Q_{Tx} Q_A + 1}{(k_{A\text{-}Rx}^2 Q_A Q_{Rx} + 1)(k_{A\text{-}Rx}^2 Q_A Q_{Rx} + k_{Tx\text{-}A}^2 Q_{Tx} Q_A + 1)}}$$

(5.21)

Therefore, in the 3-coil link, we should optimize two variables, the position of the additional resonator and Q_L. On one hand, if we want to find the optimal position of the additional resonator, we need to know the value of Q_L. On the other hand, if we want to calculate the optimum Q_L which achieves the MEP, $Q_{Lopt\text{-}\eta}$, we need to know the additional coil position ($k_{Tx\text{-}A}$ and $k_{A\text{-}Rx}$).

To optimize both variables using the equations deduced in this chapter, we should proceed as follows: First, we should find the position which maximizes $\eta_{Link\,max}$ (5.22). Second, for that position, the $Q_{Lopt\text{-}\eta}$ can be calculated from (5.21). By following these two steps, the optimum pair of Q_L and additional resonator position, which maximizes η_{Link}, is obtained. This procedure is carried out in Table 5.8, achieving a maximum $\eta_{Link} = 64\%$, which is the highest efficiency for all the combinations between additional coil position and Q_L. Following this procedure, with three identical coils, the optimum position is achieved when the additional resonator is in the middle between Rx and Tx (Fig. 5.17).

Table 5.8 Design example of 3-coil link with matching network. The calculation script is available in the supplementary material, file Sec580

Parameters for the example: Those presented in Table 5.7

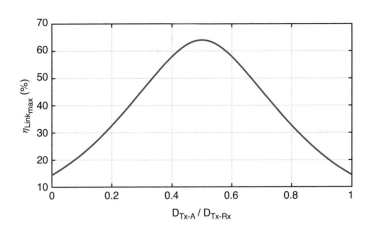

Fig. 5.17 The $\eta_{Link\,max}$ calculated from (5.22) as a function of the Tx to additional resonator distance

The optimum $D_{Tx\text{-}A} = 30\,\text{cm} \Rightarrow k_{Tx\text{-}A} = k_{A\text{-}Rx} \simeq 0.03$

at this point ($D_{Tx\text{-}A} = 30\,\text{cm}$) (5.21) $\Rightarrow Q_{Lopt\text{-}\eta} = 22$

Appendices

B.1 Deduction of $Q_{Lopt\text{-}\eta}$ Which Maximizes η_{Link}

$$\left. \begin{aligned} \eta_{Link} &= \frac{Q_{Rx\text{-}L}}{Q_L} \frac{k_{Tx\text{-}Rx}^2 Q_{Tx} Q_{Rx\text{-}L}}{k_{Tx\text{-}Rx}^2 Q_{Tx} Q_{Rx\text{-}L}+1} \quad (2.13) \\ Q_{Rx\text{-}L} &= \frac{Q_{Rx} Q_L}{Q_{Rx}+Q_L} \quad (2.11) \end{aligned} \right\} \Rightarrow$$

$$\eta_{Link} = \frac{Q_{Rx} Q_L}{(Q_{Rx}+Q_L)} \frac{1}{Q_L} \frac{k_{Tx\text{-}Rx}^2 Q_{Tx} \frac{Q_{Rx} Q_L}{(Q_{Rx}+Q_L)}}{\left(k_{Tx\text{-}Rx}^2 Q_{Tx} \frac{Q_{Rx} Q_L}{(Q_{Rx}+Q_L)} + 1 \right)} \Rightarrow$$

$$\eta_{Link} = \frac{Q_{Rx}}{(Q_{Rx}+Q_L)} \frac{k_{Tx\text{-}Rx}^2 Q_{Tx} Q_{Rx}^2 Q_L}{(k_{Tx\text{-}Rx}^2 Q_{Tx} Q_{Rx} Q_L + Q_{Rx} + Q_L)} \Rightarrow$$

$$\frac{\partial \eta_{Link}}{\partial Q_L} = \frac{Q_{Rx} \begin{bmatrix} k_{Tx\text{-}Rx}^2 Q_{Tx} Q_{Rx}^2 (Q_{Rx}+Q_L)(k_{Tx\text{-}Rx}^2 Q_{Tx} Q_{Rx} Q_L + Q_{Rx} + Q_L) \dots \\ -[(k_{Tx\text{-}Rx}^2 Q_{Tx} Q_{Rx} Q_L + Q_{Rx} + Q_L) \dots \\ +(k_{Tx\text{-}Rx}^2 Q_{Tx} Q_{Rx} + 1)(Q_{Rx}+Q_L)]k_{Tx\text{-}Rx}^2 Q_{Tx} Q_{Rx}^2 Q_L \end{bmatrix}}{(Q_{Rx}+Q_L)^2 (k_{Tx\text{-}Rx}^2 Q_{Tx} Q_{Rx} Q_L + Q_{Rx} + Q_L)^2}$$

$$\frac{\partial \eta_{Link}}{\partial Q_L} = \frac{Q_{Rx} \begin{bmatrix} \cancel{k_{Tx\text{-}Rx}^2 Q_{Tx} Q_{Rx}^2 \cdot Q_{Rx} \cdot k_{Tx\text{-}Rx}^2 Q_{Tx} Q_{Rx} Q_L} + \dots \\ k_{Tx\text{-}Rx}^2 Q_{Tx} Q_{Rx}^4 + \cancel{k_{Tx\text{-}Rx}^2 Q_{Tx} Q_{Rx}^3 Q_L} + \dots \\ \cancel{k_{Tx\text{-}Rx}^2 Q_{Tx} Q_{Rx}^2 Q_L (k_{Tx\text{-}Rx}^2 Q_{Tx} Q_{Rx} Q_L + Q_{Rx} + Q_L)} + \dots \\ - \cancel{(k_{Tx\text{-}Rx}^2 Q_{Tx} Q_{Rx} Q_L + Q_{Rx} + Q_L)k_{Tx\text{-}Rx}^2 Q_{Tx} Q_{Rx}^2 Q_L} + \dots \\ - \cancel{k_{Tx\text{-}Rx}^2 Q_{Tx} Q_{Rx} \cdot Q_{Rx} \cdot k_{Tx\text{-}Rx}^2 Q_{Tx} Q_{Rx}^2 Q_L} + \dots \\ -\cancel{1 \cdot Q_{Rx} \cdot k_{Tx\text{-}Rx}^2 Q_{Tx} Q_{Rx}^2 Q_L} + \dots \\ -(k_{Tx\text{-}Rx}^2 Q_{Tx} Q_{Rx} + 1) \cdot Q_L \cdot k_{Tx\text{-}Rx}^2 Q_{Tx} Q_{Rx}^2 Q_L \end{bmatrix}}{(Q_{Rx}+Q_L)^2 (k_{Tx\text{-}Rx}^2 Q_{Tx} Q_{Rx} Q_L + Q_{Rx} + Q_L)^2}$$

$$\frac{\partial \eta_{Link}}{\partial Q_L} = \frac{k_{Tx\text{-}Rx}^2 Q_{Tx} Q_{Rx}^3 [Q_{Rx}^2 - Q_L^2 (k_{Tx\text{-}Rx}^2 Q_{Tx} Q_{Rx} + 1)]}{(Q_{Rx}+Q_L)^2 (k_{Tx\text{-}Rx}^2 Q_{Tx} Q_{Rx} Q_L + Q_{Rx} + Q_L)^2}$$

$$\frac{\partial \eta_{Link}}{\partial Q_L} = 0 \iff Q_{Rx}^2 - Q_L^2 (k_{Tx\text{-}Rx}^2 Q_{Tx} Q_{Rx} + 1) = 0$$

$$\frac{\partial \eta_{Link}}{\partial Q_L} = 0 \iff Q_L = \frac{Q_{Rx}}{\sqrt{k_{Tx\text{-}Rx}^2 Q_{Tx} Q_{Rx} + 1}} = Q_{Lopt\text{-}\eta}$$

B.2   Deduction of $\eta_{Link\max}$

$$\left. \begin{aligned} \eta_{Link} &= \frac{Q_{Rx}}{(Q_{Rx} + Q_L)} \frac{k_{Tx\text{-}Rx}^2 Q_{Tx} Q_{Rx} Q_L}{(k_{Tx\text{-}Rx}^2 Q_{Tx} Q_{Rx} Q_L + Q_{Rx} + Q_L)} \quad (5.1) \\ Q_{Lopt\text{-}\eta} &= \frac{Q_{Rx}}{\sqrt{k_{Tx\text{-}Rx}^2 Q_{Tx} Q_{Rx} + 1}} \quad (5.3) \end{aligned} \right\} \Rightarrow$$

$$\eta_{Link\max} = \frac{Q_{Rx}}{\left(Q_{Rx} + \dfrac{Q_{Rx}}{\sqrt{k_{Tx\text{-}Rx}^2 Q_{Tx} Q_{Rx} + 1}} \right)} \cdot$$

$$\frac{k_{Tx\text{-}Rx}^2 Q_{Tx} Q_{Rx} \dfrac{Q_{Rx}}{\sqrt{k_{Tx\text{-}Rx}^2 Q_{Tx} Q_{Rx} + 1}}}{\left(k_{Tx\text{-}Rx}^2 Q_{Tx} Q_{Rx} \dfrac{Q_{Rx}}{\sqrt{k_{Tx\text{-}Rx}^2 Q_{Tx} Q_{Rx} + 1}} + Q_{Rx} + \dfrac{Q_{Rx}}{\sqrt{k_{Tx\text{-}Rx}^2 Q_{Tx} Q_{Rx} + 1}} \right)}$$

$$= \frac{Q_{Rx}}{\left(Q_{Rx} + \dfrac{Q_{Rx}}{\sqrt{k_{Tx\text{-}Rx}^2 Q_{Tx} Q_{Rx} + 1}} \right)} \frac{1}{\sqrt{k_{Tx\text{-}Rx}^2 Q_{Tx} Q_{Rx} + 1}} \cdot$$

$$\frac{k_{Tx\text{-}Rx}^2 Q_{Tx} Q_{Rx} Q_{Rx}}{\left(k_{Tx\text{-}Rx}^2 Q_{Tx} Q_{Rx} \dfrac{Q_{Rx}}{\sqrt{k_{Tx\text{-}Rx}^2 Q_{Tx} Q_{Rx} + 1}} + Q_{Rx} + \dfrac{Q_{Rx}}{\sqrt{k_{Tx\text{-}Rx}^2 Q_{Tx} Q_{Rx} + 1}} \right)}$$

$$= \frac{\cancel{Q_{Rx}}}{(\cancel{Q_{Rx}} \sqrt{k_{Tx\text{-}Rx}^2 Q_{Tx} Q_{Rx} + 1} + \cancel{Q_{Rx}})} \frac{k_{Tx\text{-}Rx}^2 Q_{Tx} Q_{Rx} \cancel{Q_{Rx}}}{\left(\dfrac{k_{Tx\text{-}Rx}^2 Q_{Tx} Q_{Rx} + 1}{\sqrt{k_{Tx\text{-}Rx}^2 Q_{Tx} Q_{Rx} + 1}} \cancel{Q_{Rx}} + \cancel{Q_{Rx}} \right)}$$

$$= \frac{1}{(\sqrt{k_{Tx\text{-}Rx}^2 Q_{Tx} Q_{Rx} + 1} + 1)} \frac{k_{Tx\text{-}Rx}^2 Q_{Tx} Q_{Rx}}{(\sqrt{k_{Tx\text{-}Rx}^2 Q_{Tx} Q_{Rx} + 1} + 1)}$$

$$= \frac{k_{Tx\text{-}Rx}^2 Q_{Tx} Q_{Rx}}{(\sqrt{k_{Tx\text{-}Rx}^2 Q_{Tx} Q_{Rx} + 1} + 1)^2}$$

B.3   Deduction of $Q_{Lopt\text{-}P_{MN}}$ (Voltage Source Tx with a Series Resonant Capacitor)

$$\left.\begin{array}{l} P_{MN} = \dfrac{V_S^2}{2R_{Tx}} \dfrac{Q_{Rx\text{-}L}}{Q_L} \dfrac{k_{Tx\text{-}Rx}^2 Q_{Tx} Q_{Rx\text{-}L}}{(k_{Tx\text{-}Rx}^2 Q_{Tx} Q_{Rx\text{-}L}+1)^2} \quad (2.17) \\[2mm] Q_{Rx\text{-}L} = \dfrac{Q_{Rx} Q_L}{Q_{Rx}+Q_L} \quad (2.11) \end{array}\right\} \Rightarrow$$

$$P_{MN} = \frac{V_S^2}{2R_{Tx}} \frac{Q_{Rx} \cancel{Q_L}}{(Q_{Rx}+Q_L)} \frac{1}{\cancel{Q_L}} \frac{k_{Tx\text{-}Rx}^2 Q_{Tx} \frac{Q_{Rx} Q_L}{(Q_{Rx}+Q_L)}}{\left(k_{Tx\text{-}Rx}^2 Q_{Tx} \frac{Q_{Rx} Q_L}{(Q_{Rx}+Q_L)} + 1\right)^2} \Rightarrow$$

$$P_{MN} = \frac{V_S^2}{2R_{Tx}} \frac{k_{Tx\text{-}Rx}^2 Q_{Tx} Q_{Rx}^2 Q_L}{(k_{Tx\text{-}Rx}^2 Q_{Tx} Q_{Rx} Q_L + Q_{Rx} + Q_L)^2} \Rightarrow$$

$$\frac{\partial P_{MN}}{\partial Q_L} = \frac{V_S^2}{2R_{Tx}} \frac{\left[\begin{array}{l} k_{Tx\text{-}Rx}^2 Q_{Tx} Q_{Rx}^2 (k_{Tx\text{-}Rx}^2 Q_{Tx} Q_{Rx} Q_L + Q_{Rx} + Q_L)^2 \ldots \\[1mm] -2(k_{Tx\text{-}Rx}^2 Q_{Tx} Q_{Rx} Q_L + Q_{Rx} + Q_L)(k_{Tx\text{-}Rx}^2 Q_{Tx} Q_{Rx} + 1)k_{Tx\text{-}Rx}^2 Q_{Tx} Q_{Rx}^2 Q_L \end{array}\right]}{(k_{Tx\text{-}Rx}^2 Q_{Tx} Q_{Rx} Q_L + Q_{Rx} + Q_L)^4}$$

$$\frac{\partial P_{MN}}{\partial Q_L} = 0 \Longleftrightarrow \begin{array}{l} (k_{Tx\text{-}Rx}^2 Q_{Tx} Q_{Rx} Q_L + Q_{Rx} + Q_L)^2 \ldots \\[1mm] -2(k_{Tx\text{-}Rx}^2 Q_{Tx} Q_{Rx} Q_L + Q_{Rx} + Q_L)(k_{Tx\text{-}Rx}^2 Q_{Tx} Q_{Rx} + 1)Q_L = 0 \end{array}$$

$$\Longleftrightarrow \begin{array}{l} (k_{Tx\text{-}Rx}^2 Q_{Tx} Q_{Rx} Q_L)^2 + (Q_{Rx} + Q_L)^2 + 2\cancel{(k_{Tx\text{-}Rx}^2 Q_{Tx} Q_{Rx} Q_L)(Q_{Rx} + Q_L)} \ldots \\[1mm] -2[(k_{Tx\text{-}Rx}^2 Q_{Tx} Q_{Rx} Q_L)^2 + \cancel{(Q_{Rx} + Q_L)(k_{Tx\text{-}Rx}^2 Q_{Tx} Q_{Rx} Q_L)} \ldots \\[1mm] +k_{Tx\text{-}Rx}^2 Q_{Tx} Q_{Rx} Q_L^2 + Q_{Rx} Q_L + Q_L^2] = 0 \end{array}$$

$$\Longleftrightarrow \begin{array}{l} -(k_{Tx\text{-}Rx}^2 Q_{Tx} Q_{Rx} Q_L)^2 + Q_{Rx}^2 + Q_L^2 + 2\cancel{Q_{Rx} Q_L} \ldots \\[1mm] -2k_{Tx\text{-}Rx}^2 Q_{Tx} Q_{Rx} Q_L^2 - 2\cancel{Q_{Rx} Q_L} - 2Q_L^2 = 0 \end{array}$$

$$\Longleftrightarrow Q_{Rx}^2 - Q_L^2 - k_{Tx\text{-}Rx}^2 Q_{Tx} Q_{Rx} Q_L^2 (k_{Tx\text{-}Rx}^2 Q_{Tx} Q_{Rx} + 2) = 0$$

$$\Longleftrightarrow Q_{Rx}^2 - Q_L^2 [k_{Tx\text{-}Rx}^2 Q_{Tx} Q_{Rx} (k_{Tx\text{-}Rx}^2 Q_{Tx} Q_{Rx} + 2) + 1] = 0$$

$$\Longleftrightarrow Q_{Rx}^2 - Q_L^2 (k_{Tx\text{-}Rx}^2 Q_{Tx} Q_{Rx} + 1)^2 = 0$$

$$\frac{\partial P_{MN}}{\partial Q_L} = 0 \Longleftrightarrow Q_L = \frac{Q_{Rx}}{k_{Tx\text{-}Rx}^2 Q_{Tx} Q_{Rx} + 1} = Q_{Lopt\text{-}P_{MN}}$$

B.4 Deduction of $P_{MN_{max}}$ (Voltage Source Tx with a Series Resonant Capacitor)

Evaluating P_{MN} at $Q_{Lopt\text{-}P_{MN}}$

$$\left.\begin{aligned}
P_{MN} &= \frac{V_S^2}{2R_{Tx}} \frac{k_{Tx\text{-}Rx}^2 Q_{Tx} Q_{Rx}{}^2 Q_L}{(k_{Tx\text{-}Rx}^2 Q_{Tx} Q_{Rx} Q_L + Q_{Rx} + Q_L)^2} \quad (5.7)\\
Q_{Lopt\text{-}P_{MN}} &= \frac{Q_{Rx}}{k_{Tx\text{-}Rx}^2 Q_{Tx} Q_{Rx} + 1} \quad\quad\quad (5.8)
\end{aligned}\right\} \Rightarrow$$

$$P_{MN} = \frac{V_S^2}{2R_{Tx}} \frac{k_{Tx\text{-}Rx}^2 Q_{Tx} Q_{Rx}{}^2 \dfrac{Q_{Rx}}{k_{Tx\text{-}Rx}^2 Q_{Tx} Q_{Rx} + 1}}{\left(k_{Tx\text{-}Rx}^2 Q_{Tx} Q_{Rx} \dfrac{Q_{Rx}}{k_{Tx\text{-}Rx}^2 Q_{Tx} Q_{Rx} + 1} + Q_{Rx} + \dfrac{Q_{Rx}}{k_{Tx\text{-}Rx}^2 Q_{Tx} Q_{Rx} + 1}\right)^2}$$

$$= \frac{V_S^2}{2R_{Tx}} \frac{k_{Tx\text{-}Rx}^2 Q_{Tx} \cancel{Q_{Rx}}{}^2 Q_{Rx}(k_{Tx\text{-}Rx}^2 Q_{Tx} Q_{Rx} + 1)}{[k_{Tx\text{-}Rx}^2 Q_{Tx} Q_{Rx} \cancel{Q_{Rx}} + \cancel{Q_{Rx}}(k_{Tx\text{-}Rx}^2 Q_{Tx} Q_{Rx} + 1) + \cancel{Q_{Rx}}]^2}$$

$$= \frac{V_S^2}{2R_{Tx}} \frac{k_{Tx\text{-}Rx}^2 Q_{Tx} Q_{Rx}(k_{Tx\text{-}Rx}^2 Q_{Tx} Q_{Rx} + 1)}{[(k_{Tx\text{-}Rx}^2 Q_{Tx} Q_{Rx} + 1) + (k_{Tx\text{-}Rx}^2 Q_{Tx} Q_{Rx} + 1)]^2}$$

$$= \frac{V_S^2}{8R_{Tx}} k_{Tx\text{-}Rx}^2 Q_{Tx} Q_{Rx} \frac{\cancel{(k_{Tx\text{-}Rx}^2 Q_{Tx} Q_{Rx} + 1)}}{\cancel{(k_{Tx\text{-}Rx}^2 Q_{Tx} Q_{Rx} + 1)^2}}$$

$$= \frac{V_S^2}{8R_{Tx}} \frac{k_{Tx\text{-}Rx}^2 Q_{Tx} Q_{Rx}}{(k_{Tx\text{-}Rx}^2 Q_{Tx} Q_{Rx} + 1)}$$

B.5   Deduction of $Q_{Lopt\text{-}P_{MN}}$ (Current Tx Source with a Series Resonant Capacitor)

$$\left.\begin{array}{l} P_{MN} = \frac{I_S^2}{2} R_{Tx} \frac{Q_{Rx\text{-}L}}{Q_L} k_{Tx\text{-}Rx}^2 Q_{Tx} Q_{Rx\text{-}L} \quad (2.19) \\ Q_{Rx\text{-}L} = \frac{Q_{Rx} Q_L}{Q_{Rx}+Q_L} \quad (2.11) \end{array}\right\} \Rightarrow$$

$$P_{MN} = \frac{I_S^2}{2} R_{Tx} \frac{Q_{Rx}}{Q_{Rx} + Q_L} k_{Tx\text{-}Rx}^2 Q_{Tx} \frac{Q_{Rx} Q_L}{Q_{Rx} + Q_L} \Rightarrow$$

$$\frac{\partial P_{MN}}{\partial Q_L} = \frac{I_S^2}{2} R_{Tx} k_{Tx\text{-}Rx}^2 Q_{Tx} Q_{Rx}^2 \frac{\partial}{\partial Q_L}\left(\frac{Q_L}{(Q_{Rx} + Q_L)^2}\right)$$

$$= \frac{I_S^2}{2} R_{Tx} k_{Tx\text{-}Rx}^2 Q_{Tx} Q_{Rx}^2 \frac{(Q_{Rx} + Q_L)^2 - 2(Q_{Rx} + Q_L)Q_L}{(Q_{Rx} + Q_L)^4}$$

$$= \frac{I_S^2}{2} R_{Tx} k_{Tx\text{-}Rx}^2 Q_{Tx} Q_{Rx}^2 \frac{Q_{Rx}^2 + Q_L^2 + 2Q_{Rx}Q_L - 2Q_{Rx}Q_L - 2Q_L^2}{(Q_{Rx} + Q_L)^4}$$

$$= \frac{I_S^2}{2} R_{Tx} k_{Tx\text{-}Rx}^2 Q_{Tx} Q_{Rx}^2 \frac{Q_{Rx}^2 - Q_L^2}{(Q_{Rx} + Q_L)^4}$$

$$\frac{\partial P_{MN}}{\partial Q_L} = 0 \Longleftrightarrow Q_L = Q_{Rx} = Q_{Lopt\text{-}P_{MN}}$$

B.6   Deduction of $P_{MN_{\max}}$ (Current Tx Source with a Series Resonant Capacitor)

$$\left.\begin{array}{l} P_{MN} = \frac{I_S^2}{2} R_{Tx} \frac{Q_{Rx\text{-}L}}{Q_L} k_{Tx\text{-}Rx}^2 Q_{Tx} Q_{Rx\text{-}L} \quad (2.19) \\ Q_{Rx\text{-}L} = \frac{Q_{Rx} Q_L}{Q_{Rx}+Q_L} \quad (2.11) \end{array}\right\} \Rightarrow$$

$$\left.\begin{array}{l} \Rightarrow P_{MN} = \frac{I_S^2}{2} R_{Tx} \frac{Q_{Rx}}{Q_{Rx}+Q_L} k_{Tx\text{-}Rx}^2 Q_{Tx} \frac{Q_{Rx} Q_L}{Q_{Rx}+Q_L} \\ Q_{Lopt\text{-}P_{MN}} = Q_{Rx} \quad (5.15) \end{array}\right\} \Rightarrow$$

$$P_{MN\max} = \frac{I_S^2}{8} R_{Tx} k_{Tx\text{-}Rx}^2 Q_{Tx} Q_{Rx}$$

B.7   Deduction of $Q_{Lopt\text{-}\eta}$ Which Maximizes η_{Link} in a 3-Coil Link

$$\left.\eta_{Link} = \frac{Q_{Rx\text{-}L}}{Q_L}\frac{k_{A\text{-}Rx}^2 Q_A Q_{Rx\text{-}L}}{(k_{A\text{-}Rx}^2 Q_A Q_{Rx\text{-}L}+1)}\frac{k_{Tx\text{-}A}^2 Q_{Tx} Q_A}{(k_{Tx\text{-}A}^2 Q_{Tx} Q_A+k_{A\text{-}Rx}^2 Q_A Q_{Rx\text{-}L}+1)}\quad (2.32)\atop Q_{Rx\text{-}L} = \frac{Q_{Rx} Q_L}{Q_{Rx}+Q_L}\qquad\qquad\qquad (2.11)\right\}\Rightarrow$$

$$\eta_{Link} = \frac{Q_{Rx}}{(Q_{Rx}+Q_L)}\frac{k_{A\text{-}Rx}^2 Q_A \frac{Q_{Rx} Q_L}{Q_{Rx}+Q_L}}{\left(k_{A\text{-}Rx}^2 Q_A \frac{Q_{Rx} Q_L}{Q_{Rx}+Q_L}+1\right)}\times\dots$$
$$\times\frac{k_{Tx\text{-}A}^2 Q_{Tx} Q_A}{\left(k_{Tx\text{-}A}^2 Q_{Tx} Q_A + k_{A\text{-}Rx}^2 Q_A \frac{Q_{Rx} Q_L}{Q_{Rx}+Q_L}+1\right)}$$

$$\eta_{Link} = Q_{Rx}\frac{k_{A\text{-}Rx}^2 Q_A Q_{Rx} Q_L}{(k_{A\text{-}Rx}^2 Q_A Q_{Rx} Q_L + Q_{Rx}+Q_L)}\times\dots$$
$$\times\frac{k_{Tx\text{-}A}^2 Q_{Tx} Q_A}{(k_{Tx\text{-}A}^2 Q_{Tx} Q_A + 1)(Q_{Rx}+Q_L) + k_{A\text{-}Rx}^2 Q_A Q_{Rx} Q_L} \tag{5.34}$$

Defining: $\mathbb{K}_{A\text{-}Rx} = k_{A\text{-}Rx}^2 Q_A Q_{Rx}$ and $\mathbb{K}_{Tx\text{-}A} = k_{Tx\text{-}A}^2 Q_{Tx} Q_A$

$$Q_{Rx}\mathbb{K}_{A\text{-}Rx}\mathbb{K}_{Tx\text{-}A}\times\dots$$

$$\frac{\partial\eta_{Link}}{\partial Q_L} = \frac{\left(\begin{array}{c}(\mathbb{K}_{A\text{-}Rx} Q_L + Q_{Rx}+Q_L)[(\mathbb{K}_{Tx\text{-}A}+1)(Q_{Rx}+Q_L)+\mathbb{K}_{A\text{-}Rx} Q_L]\\ -Q_L\left(\begin{array}{c}(\mathbb{K}_{A\text{-}Rx}+1)[(\mathbb{K}_{Tx\text{-}A}+1)(Q_{Rx}+Q_L)+\mathbb{K}_{A\text{-}Rx} Q_L]+\dots\\ +(\mathbb{K}_{A\text{-}Rx} Q_L + Q_{Rx}+Q_L)(\mathbb{K}_{A\text{-}Rx}+\mathbb{K}_{Tx\text{-}A}+1)\end{array}\right)\end{array}\right)}{\{(\mathbb{K}_{A\text{-}Rx} Q_L + Q_{Rx}+Q_L)[(\mathbb{K}_{Tx\text{-}A}+1)(Q_{Rx}+Q_L)+\mathbb{K}_{A\text{-}Rx} Q_L]\}^2}$$

$$\frac{\partial \eta_{Link}}{\partial Q_L} = 0 \Leftrightarrow \begin{array}{l} (\mathbb{K}_{A\text{-}Rx} Q_L + Q_L)[(\mathbb{K}_{Tx\text{-}A} + 1)(Q_{Rx} + Q_L) + \mathbb{K}_{A\text{-}Rx} Q_L]\ldots \\[2mm] + Q_{Rx}[(\mathbb{K}_{Tx\text{-}A} + 1)(Q_{Rx} + Q_L) + \mathbb{K}_{A\text{-}Rx} Q_L]\ldots \\[2mm] - Q_L(\mathbb{K}_{A\text{-}Rx} + 1)[(\mathbb{K}_{Tx\text{-}A} + 1)(Q_{Rx} + Q_L) + \mathbb{K}_{A\text{-}Rx} Q_L]\ldots \\[2mm] - Q_L(\mathbb{K}_{A\text{-}Rx} Q_L + Q_{Rx} + Q_L)(\mathbb{K}_{A\text{-}Rx} + \mathbb{K}_{Tx\text{-}A} + 1) = 0 \end{array}$$

$$\frac{\partial \eta_{Link}}{\partial Q_L} = 0 \Leftrightarrow \begin{array}{l} Q_{Rx}{}^2(\mathbb{K}_{Tx\text{-}A} + 1) + Q_{Rx} Q_L(\mathbb{K}_{Tx\text{-}A} + 1) + Q_{Rx} Q_L \mathbb{K}_{A\text{-}Rx} \\[2mm] - Q_L{}^2(\mathbb{K}_{A\text{-}Rx} + 1)(\mathbb{K}_{A\text{-}Rx} + \mathbb{K}_{Tx\text{-}A} + 1) \\[2mm] - Q_L Q_{Rx}(\mathbb{K}_{A\text{-}Rx} + \mathbb{K}_{Tx\text{-}A} + 1) = 0 \end{array}$$

$$\Leftrightarrow Q_L = Q_{Rx} \sqrt{\frac{k_{Tx\text{-}A}{}^2 Q_{Tx} Q_A + 1}{(k_{A\text{-}Rx}{}^2 Q_A Q_{Rx} + 1)(k_{A\text{-}Rx}{}^2 Q_A Q_{Rx} + k_{Tx\text{-}A}{}^2 Q_{Tx} Q_A + 1)}}$$

B.8   Deduction of $\eta_{Link_{max}}$ (3-Coil)

$(5.34) \Rightarrow$

$$\eta_{Link} = \begin{array}{l} Q_{Rx} \dfrac{k_{A\text{-}Rx}{}^2 Q_A Q_{Rx} Q_L}{k_{A\text{-}Rx}{}^2 Q_A Q_{Rx} Q_L + Q_{Rx} + Q_L} \times \ldots \\[4mm] \times \dfrac{k_{Tx\text{-}A}{}^2 Q_{Tx} Q_A}{(k_{Tx\text{-}A}{}^2 Q_{Tx} Q_A + 1)(Q_{Rx} + Q_L) + k_{A\text{-}Rx}{}^2 Q_A Q_{Rx} Q_L} \end{array}$$

Defining: $\mathbb{K}_{A\text{-}Rx} = k_{A\text{-}Rx}{}^2 Q_A Q_{Rx}$ and $\mathbb{K}_{Tx\text{-}A} = k_{Tx\text{-}A}{}^2 Q_{Tx} Q_A$

$$\eta_{Link} = \frac{Q_{Rx} \mathbb{K}_{A\text{-}Rx} \mathbb{K}_{Tx\text{-}A} Q_L}{(\mathbb{K}_{A\text{-}Rx} Q_L + Q_{Rx} + Q_L)[(\mathbb{K}_{Tx\text{-}A} + 1)(Q_{Rx} + Q_L) + \mathbb{K}_{A\text{-}Rx} Q_L]} \Rightarrow$$

$$\eta_{Link} = \frac{Q_{Rx} \mathbb{K}_{A\text{-}Rx} \mathbb{K}_{Tx\text{-}A} Q_L}{[Q_L(\mathbb{K}_{A\text{-}Rx} + 1) + Q_{Rx}][Q_L(\mathbb{K}_{Tx\text{-}A} + \mathbb{K}_{A\text{-}Rx} + 1) + Q_{Rx}(\mathbb{K}_{Tx\text{-}A} + 1)]}$$

dividing the numerator and denominator by $Q_{Rx} Q_L (\mathbb{K}_{Tx\text{-}A} + \mathbb{K}_{A\text{-}Rx} + 1)$

$$\eta_{Link} = \frac{\frac{\mathbb{K}_{A\text{-}Rx} \mathbb{K}_{Tx\text{-}A}}{\mathbb{K}_{Tx\text{-}A} + \mathbb{K}_{A\text{-}Rx} + 1}}{\left[\frac{Q_L}{Q_{Rx}}(\mathbb{K}_{A\text{-}Rx} + 1) + 1\right]\left[1 + \frac{Q_{Rx}(\mathbb{K}_{Tx\text{-}A} + 1)}{Q_L(\mathbb{K}_{Tx\text{-}A} + \mathbb{K}_{A\text{-}Rx} + 1)}\right]} \tag{5.35}$$

from Sect. B.7 $\Rightarrow$

$$Q_{Lopt\text{-}\eta} = Q_{Rx}\sqrt{\frac{k_{Tx\text{-}A}{}^2 Q_{Tx} Q_A + 1}{(k_{A\text{-}Rx}{}^2 Q_A Q_{Rx} + 1)(k_{A\text{-}Rx}{}^2 Q_A Q_{Rx} + k_{Tx\text{-}A}{}^2 Q_{Tx} Q_A + 1)}}$$

Using the defined: $\mathbb{K}_{A\text{-}Rx} = k_{A\text{-}Rx}{}^2 Q_A Q_{Rx}$ and $\mathbb{K}_{Tx\text{-}A} = k_{Tx\text{-}A}{}^2 Q_{Tx} Q_A$

$$Q_{Lopt\text{-}\eta} = Q_{Rx}\sqrt{\frac{\mathbb{K}_{Tx\text{-}A} + 1}{(\mathbb{K}_{A\text{-}Rx} + 1)(\mathbb{K}_{A\text{-}Rx} + \mathbb{K}_{Tx\text{-}A} + 1)}}$$

Substituting this $Q_{Lopt\text{-}\eta}$ on the η_{Link} (5.35) $\Rightarrow$

$$\eta_{Link} = \frac{\frac{\mathbb{K}_{A\text{-}Rx} \mathbb{K}_{Tx\text{-}A}}{\mathbb{K}_{Tx\text{-}A} + \mathbb{K}_{A\text{-}Rx} + 1}}{\left(\sqrt{\frac{(\mathbb{K}_{Tx\text{-}A} + 1)(\mathbb{K}_{A\text{-}Rx} + 1)}{(\mathbb{K}_{A\text{-}Rx} + \mathbb{K}_{Tx\text{-}A} + 1)}} + 1\right)\left(1 + \sqrt{\frac{(\mathbb{K}_{Tx\text{-}A} + 1)(\mathbb{K}_{A\text{-}Rx} + 1)}{(\mathbb{K}_{A\text{-}Rx} + \mathbb{K}_{Tx\text{-}A} + 1)}}\right)}$$

Defining:

$$A = \sqrt{\frac{(\mathbb{K}_{Tx\text{-}A} + 1)(\mathbb{K}_{A\text{-}Rx} + 1)}{(\mathbb{K}_{A\text{-}Rx} + \mathbb{K}_{Tx\text{-}A} + 1)}} = \sqrt{1 + \frac{\mathbb{K}_{Tx\text{-}A}\mathbb{K}_{A\text{-}Rx}}{(\mathbb{K}_{A\text{-}Rx} + \mathbb{K}_{Tx\text{-}A} + 1)}} \Rightarrow$$

$$\eta_{Link} = \frac{A^2 - 1}{(A + 1)^2}$$

B.9 Deduction of $Q_{Lopt-P_{MN}}$ (3-Coil, Voltage Source, and a Series Resonant Tx)

$$P_{MN} = \frac{V_S^2}{2R_{Tx}} \frac{Q_{Rx\text{-}L}}{Q_L} \frac{(k_{A\text{-}Rx}^2 Q_A Q_{Rx\text{-}L})(k_{Tx\text{-}A}^2 Q_{Tx} Q_A)}{(k_{Tx\text{-}A}^2 Q_{Tx} Q_A + k_{A\text{-}Rx}^2 Q_A Q_{Rx\text{-}L} + 1)^2} \quad (2.34) \Rightarrow$$

$$(2.11) \Rightarrow Q_{Rx\text{-}L} = \frac{Q_{Rx} Q_L}{Q_{Rx} + Q_L} \Rightarrow Q_L = \frac{Q_{Rx\text{-}L} Q_{Rx}}{Q_{Rx} - Q_{Rx\text{-}L}} \Rightarrow$$

$$\Rightarrow \frac{Q_{Rx\text{-}L}}{Q_L} = Q_{Rx\text{-}L} \frac{Q_{Rx} - Q_{Rx\text{-}L}}{Q_{Rx\text{-}L} Q_{Rx}} = 1 - \frac{Q_{Rx\text{-}L}}{Q_{Rx}} \Rightarrow$$

$$\frac{Q_{Rx\text{-}L}}{Q_L} = 1 - \frac{k_{A\text{-}Rx}^2 Q_A Q_{Rx\text{-}L}}{k_{A\text{-}Rx}^2 Q_A Q_{Rx}}$$

$$P_{MN} = \frac{V_S^2}{2R_{Tx}} \left(1 - \frac{k_{A\text{-}Rx}^2 Q_A Q_{Rx\text{-}L}}{k_{A\text{-}Rx}^2 Q_A Q_{Rx}}\right) \frac{(k_{A\text{-}Rx}^2 Q_A Q_{Rx\text{-}L})(k_{Tx\text{-}A}^2 Q_{Tx} Q_A)}{(k_{Tx\text{-}A}^2 Q_{Tx} Q_A + k_{A\text{-}Rx}^2 Q_A Q_{Rx\text{-}L} + 1)^2}$$

Defining: $\mathbb{K}_{A\text{-}Rx} = k_{A\text{-}Rx}^2 Q_A Q_{Rx}$, $\mathbb{K}_{Tx\text{-}A} = k_{Tx\text{-}A}^2 Q_{Tx} Q_A$ and

$$\mathbb{K}_{A\text{-}Rx\text{-}L} = k_{A\text{-}Rx}^2 Q_A Q_{Rx\text{-}L}$$

$$P_{MN} = \frac{V_S^2}{2R_{Tx}} \frac{\mathbb{K}_{A\text{-}Rx\text{-}L} \mathbb{K}_{Tx\text{-}A} \left(1 - \frac{\mathbb{K}_{A\text{-}Rx\text{-}L}}{\mathbb{K}_{A\text{-}Rx}}\right)}{(\mathbb{K}_{Tx\text{-}A} + \mathbb{K}_{A\text{-}Rx\text{-}L} + 1)^2}$$

$$\left.\begin{array}{l} \dfrac{\partial P_{MN}}{\partial Q_L} = \dfrac{\partial P_{MN}}{\partial \mathbb{K}_{A\text{-}Rx\text{-}L}} \cdot \dfrac{\partial \mathbb{K}_{A\text{-}Rx\text{-}L}}{\partial Q_L} \\[2mm] \mathbb{K}_{A\text{-}Rx\text{-}L} = k_{A\text{-}Rx}^2 Q_A \overbrace{\dfrac{Q_{Rx} Q_L}{Q_{Rx} + Q_L}}^{Q_{Rx\text{-}L}} \Rightarrow \\[2mm] \Rightarrow \dfrac{\partial \mathbb{K}_{A\text{-}Rx\text{-}L}}{\partial Q_L} = k_{A\text{-}Rx}^2 Q_A Q_{Rx} \dfrac{Q_{Rx}}{(Q_{Rx}+Q_L)^2} \end{array}\right\} \Rightarrow \dfrac{\partial \mathbb{K}_{A\text{-}Rx\text{-}L}}{\partial Q_L} \neq 0 \, \forall \, Q_L \quad \Rightarrow$$

$$\Rightarrow \frac{\partial P_{MN}}{\partial Q_L} = 0 \Leftrightarrow \frac{\partial P_{MN}}{\partial \mathbb{K}_{A\text{-}Rx\text{-}L}} = 0$$

$$\frac{\partial P_{MN}}{\partial \mathbb{K}_{A\text{-}Rx\text{-}L}} =$$

$$\frac{V_S^2}{2R_{Tx}} \frac{\left\{ \left[\mathbb{K}_{Tx\text{-}A}\left(1 - \frac{\mathbb{K}_{A\text{-}Rx\text{-}L}}{\mathbb{K}_{A\text{-}Rx}}\right) - \frac{\mathbb{K}_{A\text{-}Rx\text{-}L}\mathbb{K}_{Tx\text{-}A}}{\mathbb{K}_{A\text{-}Rx}} \right] (\mathbb{K}_{Tx\text{-}A} + \mathbb{K}_{A\text{-}Rx\text{-}L} + 1)^2 \right.}{(\mathbb{K}_{Tx\text{-}A} + \mathbb{K}_{A\text{-}Rx\text{-}L} + 1)^4}$$
$$\left. -2(\mathbb{K}_{Tx\text{-}A} + \mathbb{K}_{A\text{-}Rx\text{-}L} + 1)\left[\mathbb{K}_{A\text{-}Rx\text{-}L}\mathbb{K}_{Tx\text{-}A}\left(1 - \frac{\mathbb{K}_{A\text{-}Rx\text{-}L}}{\mathbb{K}_{A\text{-}Rx}}\right)\right] \right\}$$

$$\Rightarrow \frac{\partial P_{MN}}{\partial Q_L} = 0 \Leftrightarrow$$

$$\left\{ \left[\mathbb{K}_{Tx\text{-}A}\left(1 - \frac{\mathbb{K}_{A\text{-}Rx\text{-}L}}{\mathbb{K}_{A\text{-}Rx}}\right) - \frac{\mathbb{K}_{A\text{-}Rx\text{-}L}\mathbb{K}_{Tx\text{-}A}}{\mathbb{K}_{A\text{-}Rx}} \right] (\mathbb{K}_{Tx\text{-}A} + \mathbb{K}_{A\text{-}Rx\text{-}L} + 1)^{\not 2} \right.$$
$$\left. -2\,(\cancel{\mathbb{K}_{Tx\text{-}A} + \mathbb{K}_{A\text{-}Rx\text{-}L} + 1})\left[\mathbb{K}_{A\text{-}Rx\text{-}L}\mathbb{K}_{Tx\text{-}A}\left(1 - \frac{\mathbb{K}_{A\text{-}Rx\text{-}L}}{\mathbb{K}_{A\text{-}Rx}}\right)\right] \right\} = 0$$

$$\left\{ \cancel{\mathbb{K}_{Tx\text{-}A}}\left(1 - 2\frac{\mathbb{K}_{A\text{-}Rx\text{-}L}}{\mathbb{K}_{A\text{-}Rx}}\right)(\mathbb{K}_{Tx\text{-}A} + \mathbb{K}_{A\text{-}Rx\text{-}L} + 1) \right.$$
$$\left. -2\mathbb{K}_{A\text{-}Rx\text{-}L}\cancel{\mathbb{K}_{Tx\text{-}A}}\left(1 - \frac{\mathbb{K}_{A\text{-}Rx\text{-}L}}{\mathbb{K}_{A\text{-}Rx}}\right) \right\} = 0$$

$$\left\{ \mathbb{K}_{Tx\text{-}A} + \mathbb{K}_{A\text{-}Rx\text{-}L} + 1 - 2\frac{\mathbb{K}_{A\text{-}Rx\text{-}L}\mathbb{K}_{Tx\text{-}A}}{\mathbb{K}_{A\text{-}Rx}} - 2\frac{\cancel{\mathbb{K}_{A\text{-}Rx\text{-}L}}^2}{\mathbb{K}_{A\text{-}Rx}} - 2\frac{\mathbb{K}_{A\text{-}Rx\text{-}L}}{\mathbb{K}_{A\text{-}Rx}} \right.$$
$$\left. -2\mathbb{K}_{A\text{-}Rx\text{-}L} + 2\frac{\cancel{\mathbb{K}_{A\text{-}Rx\text{-}L}}^2}{\mathbb{K}_{A\text{-}Rx}} \right\} = 0$$

$$\left\{ \mathbb{K}_{Tx\text{-}A} + 1 + \mathbb{K}_{A\text{-}Rx\text{-}L}\left[-1 - 2\frac{\mathbb{K}_{Tx\text{-}A}}{\mathbb{K}_{A\text{-}Rx}} - \frac{2}{\mathbb{K}_{A\text{-}Rx}} \right] \right\} = 0$$

$$\Rightarrow \mathbb{K}_{A\text{-}Rx\text{-}L} = \frac{\mathbb{K}_{Tx\text{-}A} + 1}{1 + 2\frac{\mathbb{K}_{Tx\text{-}A}}{\mathbb{K}_{A\text{-}Rx}} + \frac{2}{\mathbb{K}_{A\text{-}Rx}}} = \frac{(\mathbb{K}_{Tx\text{-}A} + 1)\mathbb{K}_{A\text{-}Rx}}{\mathbb{K}_{A\text{-}Rx} + 2(\mathbb{K}_{Tx\text{-}A} + 1)} \Rightarrow$$

$$\Rightarrow \cancel{k_{A\text{-}Rx}}^2\cancel{Q_A}\frac{\overset{Q_{Rx\text{-}L}}{\overbrace{\cancel{Q_{Rx}}\,Q_L}}}{Q_{Rx} + Q_L} = \frac{(k_{Tx\text{-}A}{}^2 Q_{Tx} Q_A + 1)\cancel{k_{A\text{-}Rx}}^2\cancel{Q_A}\,Q_{Rx}}{k_{A\text{-}Rx}{}^2 Q_A Q_{Rx} + 2(k_{Tx\text{-}A}{}^2 Q_{Tx} Q_A + 1)} \Rightarrow$$

$$Q_L \left[1 - \frac{(k_{Tx\text{-}A}{}^2 Q_{Tx} Q_A + 1)}{k_{A\text{-}Rx}{}^2 Q_A Q_{Rx} + 2(k_{Tx\text{-}A}{}^2 Q_{Tx} Q_A + 1)} \right] =$$

$$Q_{Rx} \frac{(k_{Tx\text{-}A}{}^2 Q_{Tx} Q_A + 1)}{k_{A\text{-}Rx}{}^2 Q_A Q_{Rx} + 2(k_{Tx\text{-}A}{}^2 Q_{Tx} Q_A + 1)}$$

$$\Rightarrow Q_L \left[k_{A\text{-}Rx}{}^2 Q_A Q_{Rx} + k_{Tx\text{-}A}{}^2 Q_{Tx} Q_A + 1 \right] = Q_{Rx}(k_{Tx\text{-}A}{}^2 Q_{Tx} Q_A + 1)$$

$$\Rightarrow Q_L = Q_{Rx} \frac{k_{Tx\text{-}A}{}^2 Q_{Tx} Q_A + 1}{k_{Tx\text{-}A}{}^2 Q_{Tx} Q_A + k_{A\text{-}Rx}{}^2 Q_A Q_{Rx} + 1}$$

B.10 Deduction of $P_{MN_{\max}}$ (3-Coil, Voltage Source, and a Series Resonant Tx)

$$\left. \begin{aligned} P_{MN} &= \frac{V_S^2}{2R_{Tx}} \frac{Q_{Rx\text{-}L}}{Q_L} \frac{(k_{A\text{-}Rx}{}^2 Q_A Q_{Rx\text{-}L})(k_{Tx\text{-}A}{}^2 Q_{Tx} Q_A)}{(k_{Tx\text{-}A}{}^2 Q_{Tx} Q_A + k_{A\text{-}Rx}{}^2 Q_A Q_{Rx\text{-}L} + 1)^2} \quad (2.34) \\ Q_{Rx\text{-}L} &= \frac{Q_{Rx} Q_L}{Q_{Rx} + Q_L} \quad (2.11) \end{aligned} \right\} \Rightarrow$$

$$P_{MN} = \frac{V_S^2}{2R_{Tx}} \frac{Q_{Rx}}{(Q_{Rx} + Q_L)} \frac{\left(k_{A\text{-}Rx}{}^2 Q_A \frac{Q_{Rx} Q_L}{Q_{Rx} + Q_L} \right) (k_{Tx\text{-}A}{}^2 Q_{Tx} Q_A)}{\left(k_{Tx\text{-}A}{}^2 Q_{Tx} Q_A + k_{A\text{-}Rx}{}^2 Q_A \frac{Q_{Rx} Q_L}{Q_{Rx} + Q_L} + 1 \right)^2}$$

Defining: $\mathbb{K}_{A\text{-}Rx} = k_{A\text{-}Rx}{}^2 Q_A Q_{Rx}$ and $\mathbb{K}_{Tx\text{-}A} = k_{Tx\text{-}A}{}^2 Q_{Tx} Q_A$

$$\left. \begin{aligned} P_{MN} &= \frac{V_S^2}{2R_{Tx}} \frac{Q_{Rx}}{(Q_{Rx} + Q_L)} \frac{\left(\mathbb{K}_{A\text{-}Rx} \frac{Q_L}{Q_{Rx} + Q_L} \right) \mathbb{K}_{Tx\text{-}A}}{\left(\mathbb{K}_{Tx\text{-}A} + \mathbb{K}_{A\text{-}Rx} \frac{Q_L}{Q_{Rx} + Q_L} + 1 \right)^2} \\ (5.24) \Rightarrow Q_{Lopt\text{-}PMN} &= Q_{Rx} \frac{k_{Tx\text{-}A}{}^2 Q_{Tx} Q_A + 1}{k_{Tx\text{-}A}{}^2 Q_{Tx} Q_A + k_{A\text{-}Rx}{}^2 Q_A Q_{Rx} + 1} \\ &= Q_{Rx} \frac{\mathbb{K}_{Tx\text{-}A} + 1}{\mathbb{K}_{Tx\text{-}A} + \mathbb{K}_{A\text{-}Rx} + 1} \end{aligned} \right\} \Rightarrow$$

$$P_{MN\,max} =$$

$$= \frac{V_S^2}{2R_{Tx}} \frac{1}{\left(1 + \frac{\mathbb{K}_{Tx\text{-}A}+1}{\mathbb{K}_{Tx\text{-}A}+\mathbb{K}_{A\text{-}Rx}+1}\right)} \frac{\left(\mathbb{K}_{A\text{-}Rx}\left(\frac{\frac{\mathbb{K}_{Tx\text{-}A}+1}{\mathbb{K}_{Tx\text{-}A}+\mathbb{K}_{A\text{-}Rx}+1}}{\left(1+\frac{\mathbb{K}_{Tx\text{-}A}+1}{\mathbb{K}_{Tx\text{-}A}+\mathbb{K}_{A\text{-}Rx}+1}\right)}\right)\right)\mathbb{K}_{Tx\text{-}A}}{\left(\mathbb{K}_{Tx\text{-}A}+\mathbb{K}_{A\text{-}Rx}\left(\frac{\frac{\mathbb{K}_{Tx\text{-}A}+1}{\mathbb{K}_{Tx\text{-}A}+\mathbb{K}_{A\text{-}Rx}+1}}{\left(1+\frac{\mathbb{K}_{Tx\text{-}A}+1}{\mathbb{K}_{Tx\text{-}A}+\mathbb{K}_{A\text{-}Rx}+1}\right)}\right)+1\right)^2}$$

$$= \frac{V_S^2}{2R_{Tx}} \frac{\mathbb{K}_{Tx\text{-}A}+\mathbb{K}_{A\text{-}Rx}+1}{[\mathbb{K}_{A\text{-}Rx}+2(\mathbb{K}_{Tx\text{-}A}+1)]} \frac{\mathbb{K}_{A\text{-}Rx}\left(\frac{\mathbb{K}_{Tx\text{-}A}+1}{\mathbb{K}_{A\text{-}Rx}+2(\mathbb{K}_{Tx\text{-}A}+1)}\right)\mathbb{K}_{Tx\text{-}A}}{\left(\mathbb{K}_{Tx\text{-}A}+\mathbb{K}_{A\text{-}Rx}\left(\frac{\mathbb{K}_{Tx\text{-}A}+1}{\mathbb{K}_{A\text{-}Rx}+2(\mathbb{K}_{Tx\text{-}A}+1)}\right)+1\right)^2}$$

$$= \frac{V_S^2}{2R_{Tx}} \frac{\mathbb{K}_{Tx\text{-}A}+\mathbb{K}_{A\text{-}Rx}+1}{[\mathbb{K}_{A\text{-}Rx}+2(\mathbb{K}_{Tx\text{-}A}+1)]^2} \frac{\mathbb{K}_{A\text{-}Rx}(\mathbb{K}_{Tx\text{-}A}+1)\mathbb{K}_{Tx\text{-}A}}{\left(\mathbb{K}_{Tx\text{-}A}+\mathbb{K}_{A\text{-}Rx}\left(\frac{\mathbb{K}_{Tx\text{-}A}+1}{\mathbb{K}_{A\text{-}Rx}+2(\mathbb{K}_{Tx\text{-}A}+1)}\right)+1\right)^2}$$

$$= \frac{V_S^2}{2R_{Tx}} \frac{(\mathbb{K}_{Tx\text{-}A}+\mathbb{K}_{A\text{-}Rx}+1)\mathbb{K}_{A\text{-}Rx}(\mathbb{K}_{Tx\text{-}A}+1)\mathbb{K}_{Tx\text{-}A}}{\left\{[\mathbb{K}_{A\text{-}Rx}+2(\mathbb{K}_{Tx\text{-}A}+1)](\mathbb{K}_{Tx\text{-}A}+1)+\mathbb{K}_{A\text{-}Rx}(\mathbb{K}_{Tx\text{-}A}+1)\right\}^2}$$

$$= \frac{V_S^2}{2R_{Tx}} \frac{(\cancel{\mathbb{K}_{Tx\text{-}A}+\mathbb{K}_{A\text{-}Rx}+1})\mathbb{K}_{A\text{-}Rx}(\cancel{\mathbb{K}_{Tx\text{-}A}+1})\mathbb{K}_{Tx\text{-}A}}{4(\mathbb{K}_{Tx\text{-}A}+\mathbb{K}_{A\text{-}Rx}+1)^{\cancel{2}}(\mathbb{K}_{Tx\text{-}A}+1)^{\cancel{2}}} \Longrightarrow$$

$$P_{MN\,max} = \frac{V_S^2}{8R_{Tx}} \frac{k_{Tx\text{-}A}{}^2 Q_{Tx} Q_A . k_{A\text{-}Rx}{}^2 Q_A Q_{Rx}}{(k_{Tx\text{-}A}{}^2 Q_{Tx} Q_A + 1)(k_{A\text{-}Rx}{}^2 Q_A Q_{Rx} + k_{Tx\text{-}A}{}^2 Q_{Tx} Q_A + 1)}$$

References

1. M.E. Van Valkenburg, *Network Analysis* (Prentice-Hall, 1964)
2. T.H Lee, *Planar Microwave Engineering: A Practical Guide to Theory, Measurement, and Circuits*, vol. 1 (Cambridge University Press, 2004)
3. P. Pérez-Nicoli, F. Silveira, Maximum efficiency tracking in inductive power transmission using both matching networks and adjustable AC-DC converters. IEEE Trans. Microw. Theory Techn. **66**(7),3452–3462 (2018)
4. R.F. Xue, K.W. Cheng, M. Je, High-efficiency wireless power transfer for biomedical implants by optimal resonant load transformation. IEEE Trans. Circuits Syst. I **60**(4),867–874 (2013) ISSN 1549-8328. https://doi.org/10.1109/TCSI.2012.2209297
5. Y. Han, D.J. Perreault, Analysis and design of high efficiency matching networks. IEEE Trans. Power Electron. **21**(5),1484–1491 (2006)
6. M. Zargham, P.G. Gulak, Maximum achievable efficiency in near-field coupled power-transfer systems. IEEE Trans. Biomed. Circuits Syst. **6**(3),228–245 (2012)

Chapter 6
Adaptive Circuits to Track the Optimum Operating Point (OOP)

6.1 Introduction

As discussed in Chap. 5, considering the generic WPT link of Fig. 6.1, there are optimum values for the Rx coil load impedance, Z_{MN}, which maximize the link efficiency, η_{Link}, or the power received, P_{MN}, achieving the Maximum Efficiency Point (MEP) or Maximum Power Point (MPP), respectively. The Optimum Operating Point (OOP) is used to refer to both MEP and MPP, depending on the desired link optimization paradigm, which is in turn linked to the application. The optimum Z_{MN} is represented by $Z_{MN_{opt}}$ and it depends on the value of the coupling coefficient between coils and load resistance R_L. These usually change during the system operation, due to changes in the alignment and distance between the coils and changes in the consumption of the circuit that is being powered. Therefore, an effective design for the OOP requires to actively adapt to these changing conditions. This chapter presents different circuits to track the OOP. Let us summarize the main path that determines $Z_{MN_{opt}}$. The imaginary part of $Z_{MN_{opt}}$ is the one that achieves resonance by canceling the Rx coil impedance, $j\omega L_{Rx}$. The real part of $Z_{MN_{opt}}$ depends on what is being maximized, either η_{Link} or P_{MN}. Additionally, that optimum real part also depends on the number of coils in the link and the coupling coefficient between coils, among other parameters, as studied in Chap. 5.

The Rx matching network is commonly used to achieve resonance, $Im\{Z_{MN}\} = Im\{Z_{MN_{opt}}\} = -j\omega L_{Rx}$, by using a simple series or parallel resonant capacitor. In Chap. 5, the use of the Rx matching network to achieve both $Im\{Z_{MN}\} = Im\{Z_{MN_{opt}}\}$ and $Re\{Z_{MN}\} = Re\{Z_{MN_{opt}}\}$ was also addressed. In Table 5.4, some matching networks and its impedance conversions were presented.

The input impedance of a matching network depends on its load impedance, as shown in Table 5.4. The Rx matching network is loaded by the rectifier, the DC-DC converter, and the actual load, in cascade as shown in Fig. 6.1. The rectifier input resistance, R_{rect}, and DC-DC converter input resistance, $R_{DC\text{-}DC}$, are approximations, and its reactive components are neglected for the sake of simplicity.

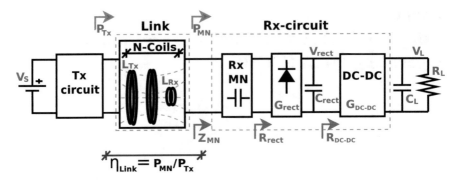

Fig. 6.1 A generic WPT link to recall the definition of Tx output power P_{Tx}, received power P_{MN}, link efficiency η_{Link}, Rx coil load impedance Z_{MN}, rectifier input resistance R_{rect}, and DC-DC input resistance $R_{DC\text{-}DC}$

Additionally, the harmonics generated by the rectifier are not considered as they are filtered by the resonant tank.

The power consumption of the circuit that is being powered by the WPT link could change, which is modeled by a change in R_L. This would affect the DC-DC input resistance, $R_{DC\text{-}DC}$, and the rectifier input resistance, R_{rect}, changing Z_{MN}. Therefore, it is not trivial to keep $Z_{MN} = Z_{MN_{opt}}$ if R_L changes during operation. Moreover, $Z_{MN_{opt}}$ could also dynamically change as it depends on the coupling factor between coils, as deduced in Chap. 5.

Therefore, in order to keep the OOP condition,

$$Z_{MN} = Z_{MN_{opt}}$$

depends on load,   depends on coupling,
R_L, variations k, variations

(6.1)

under loading, R_L, and coupling, k, variations, it is necessary to use an adaptive circuit that reacts to those variations.

Different adaptive circuits used to hold the OOP under R_L and k variations are addressed in this chapter. First, in Sects. 6.2.1 and 6.2.2, the use of switched-inductor and switched-capacitor DC-DC converters is addressed, respectively. Then, in Sect. 6.3 it is analyzed how active rectifiers could also be used for OOP tracking. In Sect. 6.4, other approaches that are implemented in the AC domain, i.e., before the rectifier, are presented. Those other approaches are (1) Q-modulation, Sect. 6.4.1; (2) adaptive Rx matching network, Sect. 6.4.2; and (3) reconfigurable Rx coils, Sect. 6.4.3. Finally, in Sect. 6.5, the joint use of fixed (nonadaptive) Rx matching network and the adaptive circuits, addressed in this chapter, is discussed.

Before starting, it is worth noting that the same adaptive circuits that will be introduced in this chapter to track the OOP could be instead used for output voltage regulation. In this chapter, we focus on the OOP tracking, without considering output voltage regulation. The design to simultaneously achieve tracking of the OOP and regulation of the output voltage is addressed in Chap. 7.

6.2 Using the Rx DC-DC Converter to Achieve the OOP

As discussed in the introduction of this chapter, Z_{MN} depends on the matching network load resistance, R_{rect}, which in turn could be affected by the rectifier and DC-DC converter. The relationship between Z_{MN} and R_{rect} depends on the matching network topology used, and it was addressed in Table 5.4. The relationship between R_{rect} and $R_{DC\text{-}DC}$ will be further addressed in Sect. 6.3, but basically, R_{rect} is proportional to $R_{DC\text{-}DC}$, $R_{rect} \propto R_{DC\text{-}DC}$. Finally, the relationship between $R_{DC\text{-}DC}$ and R_L is addressed in this section for the switched-inductor (Sect. 6.2.1) and switched-capacitor (Sect. 6.2.2) topologies. Therefore, by changing $R_{DC\text{-}DC}$, the DC-DC converter can be used to adjust Z_{MN} achieving the OOP, $Z_{MN} = Z_{MN_{opt}}$.

6.2.1 Switched-Inductor Converters

The three main topologies of switched-inductor DC-DC converters are presented in Fig. 6.2. The load capacitance, C_L, is assumed to be large enough so that the output voltage, V_L, is constant. In the three cases, the switch is turned ON and OFF with certain frequency, f_{sw}, and duty cycle, D_{sw}, graphically represented in Fig. 6.2a, which impact the converter gain V_L/V_{rect}. The buck converter (Fig. 6.2b) steps down the input voltage ($V_L < V_{rect}$), while the boost converter (Fig. 6.2c) steps it up ($V_L > V_{rect}$). The buck-boost topology (Fig. 6.2d) has an output voltage that can be either greater or less than the input voltage.

A detailed analysis of switched-inductor converters can be found in classical references, such as [1]. In this section, we summarize its use to modify Z_{MN}.

The operation mode of a switched-inductor DC-DC converter can be divided into Continuous Conduction Mode (CCM) and Discontinuous Conduction Mode (DCM). When the current through the inductor, L_{sw}, is always greater than zero, the converter is operating in CCM. However, when the current through the inductor becomes zero during operation, the converter is said to be operating in DCM.

When the converter is operating in CCM, the gain and relationship between input and output resistances are independent of the switching frequency, f_{sw}, and only depend on the duty cycle, D_{sw}. The simplified expressions for gain and input resistance for the buck, boost, and buck-boost converters, operating in CCM and neglecting the diode voltage drop, are summarized in Table 6.1. Although the input current of switched converters is not constant, it is filtered by the rectifier output capacitance, C_{rect}. Therefore, the switched converter input resistance can be defined as the ratio between the input voltage and the average input current.

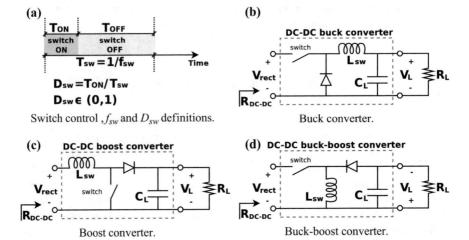

Fig. 6.2 Three main topologies of switched-inductor DC-DC converter: buck (**b**), boost (**c**), and buck-boost (**d**). The switching frequency, f_{sw}, and duty cycle, D_{sw}, are defined in (**a**)

Table 6.1 Gain and input resistance for switched-inductor DC-DC converters operating in Continuous Conduction Mode (CCM)

Topology	Gain	Input resistance $R_{DC\text{-}DC} = \dfrac{V_{rect}}{\text{average input current}}$
Buck	$V_L/V_{rect} = D_{sw}$	$R_{DC\text{-}DC} = \dfrac{1}{D_{sw}^2} R_L$
Boost	$V_L/V_{rect} = \dfrac{1}{1-D_{sw}}$	$R_{DC\text{-}DC} = (1 - D_{sw})^2 R_L$
Buck-boost	$V_L/V_{rect} = \dfrac{D_{sw}}{1-D_{sw}}$	$R_{DC\text{-}DC} = \left(\dfrac{1}{D_{sw}} - 1\right)^2 R_L$

Note that with a buck converter, the $R_{DC\text{-}DC}$ can only be set to values that are higher than R_L, as $D_{sw} \in (0, 1)$. On the other hand, by using a boost converter, $R_{DC\text{-}DC}$ can only be set to lower values than R_L. In theory, any desired value for $R_{DC\text{-}DC}$ can be achieved if a buck-boost converter is used. In the three topologies, operating in CCM, $R_{DC\text{-}DC}$ depends on R_L. Therefore, if R_L changes, the duty cycle, D_{sw}, should be adjusted to hold the optimum $R_{DC\text{-}DC}$ which satisfies $Z_{MN} = Z_{MN_{opt}}$. In [2], the buck-boost converter shown in Fig. 6.2d was used and its duty cycle was adjusted in closed-loop to track the OOP. Similarly, in [3], the cascade of a boost and buck converters was used with the same goal.

Unlike CCM, in DCM the gain and input resistance of the DC-DC converter are affected by f_{sw} and L_{sw}. On the one hand, the buck-boost converter operating in DCM has an input resistance, $R_{DC\text{-}DC}$, which is independent of its load resistance, R_L. In that case, $R_{DC\text{-}DC} = \dfrac{2 L_{sw} f_{sw}}{D_{sw}^2}$. Therefore, once f_{sw} and D_{sw} are selected to achieve the desired $R_{DC\text{-}DC}$, it is not affected by changes in R_L. This idea was used in [4] to achieve the OOP regardless of R_L variations. However, if the coupling coefficient changes, $Z_{MN_{opt}}$ is altered, and it is required to adjust f_{sw} or D_{sw} to maintain $Z_{MN} = Z_{MN_{opt}}$.

6.2.2 Switched-Capacitor Converters

Usually, the inductors required by switched-inductor converters cannot be integrated, preventing us from using this architecture in applications with extreme size constraints. Switched-capacitor converters, which only use capacitors and switches, are more suitable for integration and advantageous in strongly size-constrained applications such as highly miniaturized AIMDs.

The design of switched-capacitor converters is out of the scope of this book, as it can be found in classical references, such as [5]. In this section, we focus on the use of these converters to adjust the seen impedance.

The input resistance $R_{DC\text{-}DC}$ of a switched-capacitor converter can also be adjusted by modifying the switching frequency and duty cycle like in switched-inductor converters. However, its use in WPT links to track the OOP is less common than the use of switched-inductors converters. In WPT links, switched-capacitor converters are used, but typically with the aim of regulating the output voltage [6], as opposed to tracking the OOP. However, switched-capacitor converters have been used to track the Maximum Power Point (MPP), i.e., adjust its input resistance, in photovoltaic sources [7].

The efficiency of switched-capacitor converters is deteriorated when the switching frequency or duty cycle are altered, from their optimum values, either with the goal of output voltage or input resistance regulation. Therefore, reconfigurable multiple-gain architectures are used which provide wider output voltage or input resistance regulation with high efficiency [6, 8].

For the generic DC-DC converter with efficiency, $\eta_{DC\text{-}DC} = P_L/P_{DC\text{-}DC}$, and gain, $G_{DC\text{-}DC} = V_L/V_{rect}$, presented in Fig. 6.3, the input resistance can be approximated as presented in (6.2).

$$P_L = \eta_{DC\text{-}DC} P_{DC\text{-}DC} \Rightarrow$$

$$\eta_{DC\text{-}DC} = \frac{P_L}{P_{DC\text{-}DC}} = \frac{V_L{}^2/R_L}{V_{rect}{}^2/R_{DC\text{-}DC}} = \frac{R_{DC\text{-}DC}}{R_L} G_{DC\text{-}DC}{}^2 \Rightarrow$$

$$R_{DC\text{-}DC} = \frac{R_L \cdot \eta_{DC\text{-}DC}}{G_{DC\text{-}DC}{}^2}$$

(6.2)

As deduced in (6.2), $R_{DC\text{-}DC}$ can be modified by changing the converter gain. Switching between step-up and step-down topologies, in theory, any desired $R_{DC\text{-}DC}$ can be achieved (higher or lower than R_L). However, when a too low or too high gain is required, i.e., $G_{DC\text{-}DC} > 10$ or $G_{DC\text{-}DC} < 0.1$, it is difficult to achieve high efficiency in the DC-DC converter, $\eta_{DC\text{-}DC}$, and too many capacitors are required.

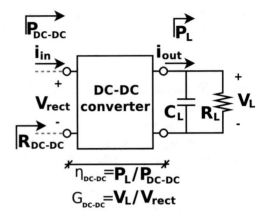

In that case, another part of the system should be modified, as will be discussed in Sect. 6.5.

6.3   Using an Active Rectifier to Achieve the OOP

The rectifier is a key building block of the WPT system. Therefore, the design of rectifiers with high Power Conversion Efficiency (PCE) and Voltage Conversion Ratio (VCR) has been frequently addressed in literature [9–15]. Passive diodes dramatically degrade PCE and VCR in applications where the input voltage amplitude is comparable with the diode forward voltage drop. To solve this, in active rectifiers, the passive diodes are substituted with MOS switches and comparators as shown in Fig. 6.4 [11–14], avoiding the voltage drop and increasing both PCE and VCR. In Fig. 6.4, when the voltage in the anode is higher than in the cathode, the comparator turns ON the MOS switch, thus conducting current. On the other hand, when the cathode has a voltage that is higher than the anode, the switch is turned OFF. This results in the active diode behaving similar to an ideal diode with zero turn-on voltage. Some disadvantages of active diodes stem from the comparator performance: its speed, offset, and the power consumption that it adds.

The active rectifier can be used to achieve higher PCE and/or VCR (compared to a passive implementation), followed by a DC-DC converter to track the OOP as shown in Sect. 6.2. However, many authors have taken advantage of the active rectifier, using it to track the OOP and avoid using the DC-DC converter. In that approach, the rectifier and DC-DC converter, shown in Fig. 6.1, are implemented together, as an AC-DC block, reducing the number of components. In this case, the rectifier input resistance can be controlled by adjusting the signals that control the switches (Sect. 6.3.1) or with a reconfigurable architecture (Sect. 6.3.2).

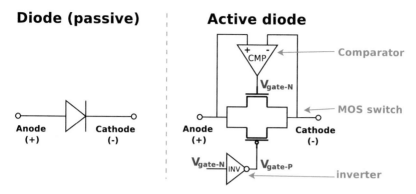

Fig. 6.4 Active diode used to avoid diode forward voltage

6.3.1   Modifying the Control Signals

Different architectures of active rectifiers have been proposed in the literature. For instance, the classic full-wave rectifier that uses four diodes in a bridge configuration (Fig. 6.5a) could be implemented with MOS switches (active diodes) (Fig. 6.5b) [16, 17]. Since each transistor has to be turned ON and OFF in each period, a considerable amount of power is used in gate drive. In the semi-active implementation [18–21] (Fig. 6.5c), only two diodes are replaced by MOS switches, reducing the number of required signals and the gate drive power consumption. It is also possible to reduce the gate drive losses connecting the PMOS transistors as shown in Fig. 6.5d, which is a widely used architecture in the literature [11, 12, 22, 23]. In the architecture presented in Fig. 6.5d, which operation is explained below, the PMOS switches are turned ON and OFF by themselves avoiding approximately half of the gate drive losses. Instead of cross-connecting the PMOS, it is also possible to cross-connect the NMOS and control the gates of PMOS switches as in [24–26].

The implementation of the cross-connected active rectifier of Fig. 6.5d is depicted in Fig. 6.6. The main voltages of Fig. 6.6 are represented as a function of time in Fig. 6.7, where four operating zones are defined, as depicted in Fig. 6.8.

The operating zones are determined by the inputs, V_1 and V_2, and the output, V_{rect}, voltages. While $0 < V_1 - V_2 < V_{rect}$ (Zone I), only MP1 is ON; thus $V_1 = V_{rect}$. Since $V_1 = V_{rect}$, the increase in the differential-input voltage ($V_1 - V_2$) generates a reduction in V_2 until it reaches ground, which happens when $V_1 - V_2 = V_{rect}$. When $V_1 - V_2 > V_{rect}$ (zone II), MN2 is turned ON, and the charge is delivered to the output node through transistors MN2 and MP1. Zones III and IV are analogous to zones I and II, respectively, but with a different polarity.

Three different approaches to control the input resistance of active rectifiers have been proposed: (1) Pulse-Width Modulation (PWM) [18, 20], (2) Pulse-Density Modulation (PDM) [27, 28], and (3) Pulse-Shift Modulation (PSM) [19]. These three approaches are graphically represented in Fig. 6.9. The PWM and PDM do not introduce a phase shift between the rectifier input voltage and current. Therefore,

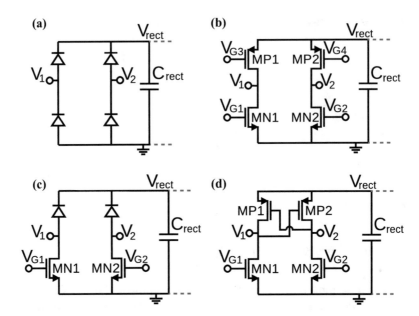

Fig. 6.5 Full-wave rectifiers in a bridge configuration: passive, semi-active, and active implementation. For the sake of simplicity, the comparators that control the gates to implement the active diodes, as was depicted in Fig. 6.4, are not shown here. (**a**) Passive. (**b**) Active. (**c**) Semi-active. (**d**) Active, cross-connected PMOS gates

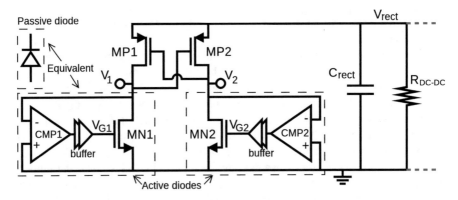

Fig. 6.6 Widely used active rectifier architecture which reduces gate drive power losses by cross-connecting the PMOS gates

PWM and PDM only affect the rectifier input resistance. However, PSM introduces a phase shift that generates (or modifies) the rectifier input reactance. It is also possible to combine the approaches presented in Fig. 6.9. In [25], the PWM and PDM are jointly used to achieve a tighter control. In [16, 21], the PWM and PSM are jointly used to control both the real and imaginary parts of the rectifier input impedance, which allow tracking the OOP and Rx resonance at the same time.

Fig. 6.7 Rectifier input voltages, V_1 and V_2, and gate control signals, V_{G1} and V_{G2}, for the circuit presented in Fig. 6.6

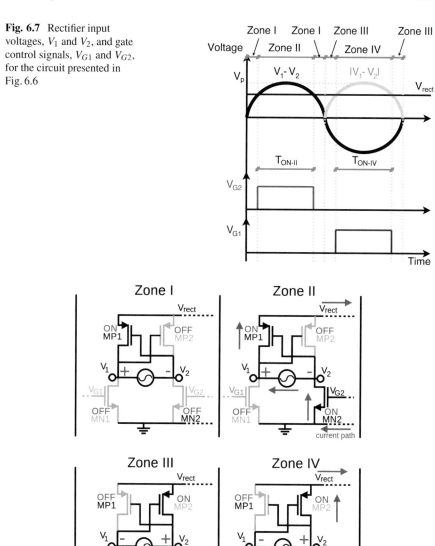

Fig. 6.8 Detailed operation of the active rectifier presented in Fig. 6.6

The active rectifier can be controlled as introduced in this section with the aim of tracking the OOP [16, 19, 28] or with the aim of output voltage regulation [18, 20, 21, 25, 27]. The design to track the OOP and regulate the output voltage at the same time will be addressed in Chap. 7.

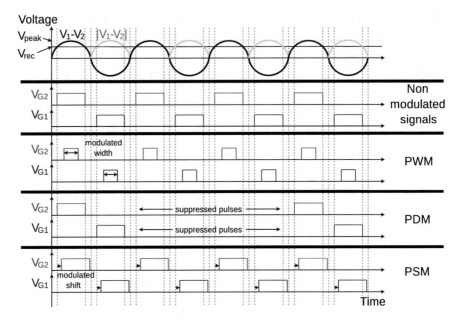

Fig. 6.9 Main approaches to adjust the active rectifier control signals

6.3.2   Reconfigurable Multiple-Gain Architectures

Instead of adjusting the control signals of the active rectifier, a reconfigurable gain architecture can be used like in the case of switched-capacitor DC-DC converters (Sect. 6.2.2). Different reconfigurable passive [29–31] and active [15, 32–34] rectifier architectures have been proposed in the literature. By way of example, we describe here the architecture proposed in [34], presented in Fig. 6.10a, which is based on the cross-connected PMOS gate topology introduced in Fig. 6.5d. The architecture of Fig. 6.10a can be use in ×1 mode (Fig. 6.10b), which is identical to Fig. 6.5d, or in ×2 mode (Fig. 6.10c), which is a voltage doubler made up of two half-wave rectifiers connected in series [35]. Neglecting the dropout voltage, the output voltage, V_{rect}, is equal to the peak differential-input voltage, $(V_1 - V_2)_{peak}$ in the ×1 mode, $V_{rect} = (V_1 - V_2)_{peak}$ at ×1, while it is twice as much in the ×2 mode, $V_{rect} = 2 \times (V_1 - V_2)_{peak}$ at ×2.

By changing the rectifier gain, the relationships between its input and output resistances are altered as follows. As was carried out for the reconfigurable DC-DC converters, it is possible to deduce a first-order approximation for R_{rect} by equating the DC output power to the first harmonic AC input power [32]. This approximation is presented in Fig. 6.11 and (6.3), where $V_{in-1_{peak}}$ is the first harmonic peak voltage at the input of the rectifier:

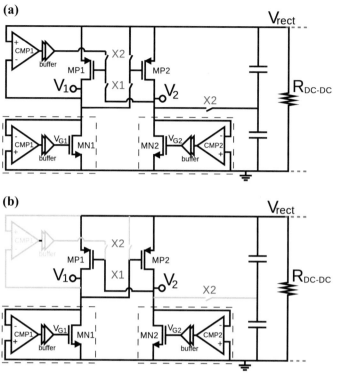

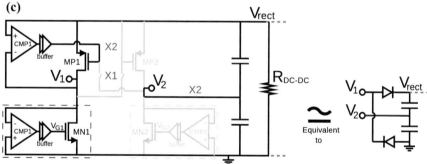

Fig. 6.10 Example of reconfigurable active rectifier. (**a**) Reconfigurable topology. (**b**) Mode ×1, equal to Fig. 6.6. (**c**) Mode ×2, voltage doubler made up of two half-wave rectifiers connected in series [35]

$$P_{DC\text{-}DC} = \eta_{rect} P_{rect} \Rightarrow$$

$$\eta_{rect} = \frac{P_{DC\text{-}DC}}{P_{rect}} = \frac{V_{rect}^2/R_{DC\text{-}DC}}{V_{in\text{-}1\,peak}^2/(2R_{rect})} = \frac{2R_{rect}}{R_{DC\text{-}DC}} G_{rect}^2 \Rightarrow \qquad (6.3)$$

$$R_{rect} = \frac{R_{DC\text{-}DC} \cdot \eta_{rect}}{2G_{rect}^2}$$

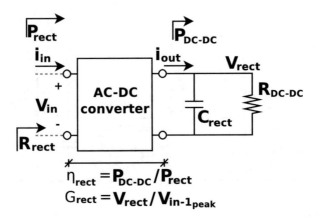

Fig. 6.11 Generic rectifier with adaptive gain to analyze the relationship of input/output resistances

For instance, operating at ×2 mode, and neglecting power losses ($\eta_{rect} = 1$), $R_{rect} = \frac{1}{8}R_{DC\text{-}DC}$ [32].

Like all the previously presented adaptive circuits, reconfigurable active rectifiers can be used to track the OOP [32] or to regulate the output voltage [33, 35]. In [33, 35], a feedback loop that automatically switches between ×1 and ×2 mode to regulate the output voltage is proposed. In [32], the control feedback loop to track the OOP is not addressed, but it is proven how the efficiency of the link is improved by adjusting the rectifier gain and thus operating closer to the OOP.

6.4   OOP Tracking in the AC Domain

6.4.1   Q-Modulation

In Q-modulation the Rx series resonator is chopped, i.e., the input of the rectifier is shorted, to adapt Z_{MN} with the aim of tracking the OOP (Fig. 6.12).

The effect that shorting the rectifier input has on Z_{MN} strongly depends on the switch control signal. Three different shorting strategies are summarized in Fig. 6.13.

In the Q-modulation technique proposed in [36], the switch is turned ON twice in every power carrier cycle in a synchronized fashion with the current through the Rx coil, i_{Rx}. The Z_{MN} is controlled by adjusting the switching duty cycle to achieve the OOP. This switching strategy requires high switching frequency and a sophisticated synchronization between i_{Rx} and the switch control signal.

These drawbacks were solved in the multi-cycle non-synchronized Q-modulation technique proposed in [37]. In this approach, the switch remains ON during multiple carrier cycles without requiring synchronization. In [37], it is proven that the delay

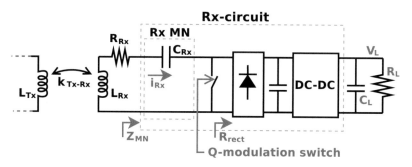

Fig. 6.12 Q-modulation technique

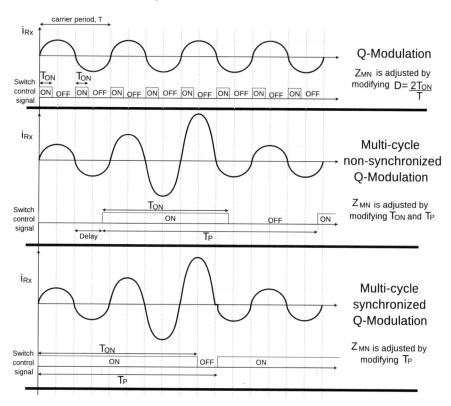

Fig. 6.13 Q-modulation shorting strategies. Q-modulation [36], multi-cycle non-synchronized Q-modulation [37], and multi-cycle synchronized Q-modulation [38]

between i_{Rx} and the switch control signal, i.e., synchronization, does not affect the system performance, and Z_{MN} is adjusted by modifying the ON time, T_{ON}, and switching period, T_P.

Another approach is presented in [38], where the switch is ON for several power carrier cycles, thus reducing to almost zero the impedance seen to the rectifier which

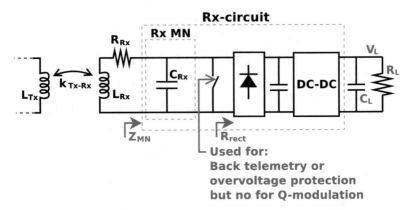

Fig. 6.14 Shorting the rectifier input in parallel compensated Rx

tends to increase the current through the Rx coil. Then, when the current is at its peak value, i.e., all the energy is stored in the Rx coil and the capacitor C_{Rx} is discharged, the switch is turned OFF to deliver the energy stored in the Rx coil to the Rx-circuit in a short period of time. This approach has similarities with a boost converter since a step-up effect is achieved by charging a coil (switching inductor). This step-up effect is able to reduce the seen resistance similar to the variable gain rectifiers addressed in Sect. 6.3.2 (see (6.3) with $G_{rect} > 1$).

The Q-modulation techniques addressed in this section are feasible when a series Rx resonant capacitor is used. As discussed in Sect. 5.6, the series resonant topology is preferred in high-power applications, where R_L is in the same order of magnitude as R_{Rx}. The effect of shorting the rectifier input in a parallel compensated Rx (Fig. 6.14) is completely different. When the switch shown in Fig. 6.14 is ON, the Rx is strongly detuned and the efficiency of the link is spoiled. Therefore, shorting the rectifier input in a parallel compensated Rx does not serve to track the OOP. As shown in Chap. 4, this architecture is used for back telemetry [9, 32, 39], because the significant change in Z_{MN} can be detected from the Tx side. Additionally, shorting the rectifier input in a parallel compensated Rx has been used in the literature for overvoltage protection [40] because detuning the Rx coil reduces the received power, P_{MN}, and the rectifier input voltage can be reduced down to almost zero.

6.4.2 Adaptive Matching Network

As discussed in Sect. 5.6, a matching network can be used to adjust Z_{MN}, achieving resonance and/or the OOP. The matching networks presented in Sect. 5.6 are fixed (non-adjustable); thus they can be used to achieve the OOP at specific conditions of coupling, k, and load power consumption, R_L.

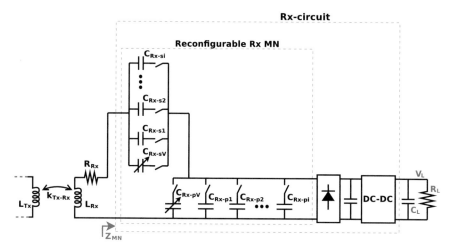

Fig. 6.15 Example of reconfigurable Rx matching network to track the OOP

Although it is not a typical approach, adaptive matching networks can be used to dynamically track the OOP. It can be implemented using varactor diodes, reconfigurable structures, or both [41], as shown in Fig. 6.15.

Using a reconfigurable matching network, the OOP can be tracked over a wide loading range. However, the implementation is complex. First, all the reconfigurable topologies should be designed to maintain the resonance, $Im\{Z_{MN}\} = -j\omega L_{Rx}$, while taking into account the parasitic capacitance of the switches, and only the real part, $Re\{Z_{MN}\}$, should be adjusted. Second, depending on the topology and power level, the switches may have to tolerate large voltages. Third, the ON-resistance of the switches deteriorates the system efficiency. Therefore, usually, the OOP is not tracked using a reconfigurable matching network. As will be shown in Sect. 6.5, in order to track the OOP, a fixed matching network could be used for a coarse adjustment, while a DC-DC converter (Sect. 6.2) or active rectifier (Sect. 6.3) achieves the fine-tuning and dynamically tracks the OOP.

Reconfigurable matching networks are more commonly used in the literature to only adjust the imaginary part $Im\{Z_{MN}\}$, in order to track resonance or detune the Rx. For instance, if changes in the environment have resulted in detuning the Rx, an auto-tuning mechanism has been proposed in [42] to increase the received power by adjusting the matching circuitry. Alternatively, in [43], the Rx is detuned to limit the received power with the aim of regulating the output voltage.

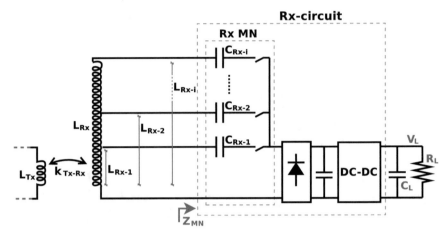

Fig. 6.16 Reconfigurable Rx coil to track the OOP

6.4.3 Reconfigurable Resonant Coil

All the previously presented strategies were focused on modifying $Re\{Z_{MN}\}$. In this approach, instead of matching $Re\{Z_{MN}\}$ to $Re\{Z_{MN_{opt}}\}$, the $Re\{Z_{MN_{opt}}\}$ is adjusted to match $Re\{Z_{MN}\}$.

To do this, the multi-tapped Rx coil presented in Fig. 6.16 is used. The capacitors are selected to resonate with its series inductance at the carrier frequency, i.e., $f = \frac{1}{2\pi\sqrt{L_{Rx\text{-}i}C_{Rx\text{-}i}}}$ $\forall i$ (where f is the carrier frequency), and only one switch is ON at a time. Therefore, by changing the ON switch in Fig. 6.16, the Rx coil is modified while maintaining resonance.

As studied in Chap. 5, the optimum value of $Re\{Z_{MN}\}$, $Re\{Z_{MN_{opt}}\}$, depends on (1) the number of coils used in the inductive link, (2) the coil quality factors, (3) the coupling coefficients between coils, and (4) the Rx coil parasitic series resistance, R_{Rx} (see the optimum values deduced in (5.4), (5.9), (5.16), (5.23) and (5.25)). By using a multi-tapped Rx coil, it is possible to modify its quality factor Q_{Rx}, coupling with the subsequent coil, e.g., $k_{Tx\text{-}Rx}$, and its parasitic series resistance, R_{Rx}. Therefore, it is possible to adjust $Re\{Z_{MN_{opt}}\}$ in order to achieve the OOP, $Re\{Z_{MN}\} = Re\{Z_{MN_{opt}}\}$.

However, it should be noted that the maximum efficiency, η_{Link}, or power, P_{MN}, obtained at the OOP is also affected by those parameters. As was studied in Sect. 2.1.5, the quality factor and coupling coefficient should be as high as possible; thus reducing them to achieve the OOP could be counterproductive. As an example, let us assume that we want to achieve the MEP of a 2-coil link. In that case, according to Chap. 5

$$Re\{Z_{MN_{opt}}\} = Re\{Z_{MN_{opt\text{-}\eta}}\} = R_{Rx}\sqrt{k_{Tx\text{-}Rx}{}^2 Q_{Tx} Q_{Rx} + 1}, \tag{5.4}$$

and the maximum efficiency achieved at this $Re\{Z_{MN_{opt-\eta}}\}$ is

$$\eta_{Link\,max} = \frac{k_{Tx\text{-}Rx}^2 Q_{Tx} Q_{Rx}}{(\sqrt{k_{Tx\text{-}Rx}^2 Q_{Tx} Q_{Rx} + 1} + 1)^2}. \tag{5.5}$$

If $k_{Tx\text{-}Rx}$ or Q_{Rx} is reduced to reach $Re\{Z_{MN}\} = Re\{Z_{MN_{opt-\eta}}\}$, $\eta_{Link\,max}$ will be deteriorated. Therefore, although the MEP is achieved, the obtained $\eta_{Link\,max}$ is lower than the one achieved by using the previously presented methods that do not affect $\eta_{Link\,max}$. Moreover, if the $k_{Tx\text{-}Rx}$ or Q_{Rx} is reduced to reach the MEP, $Re\{Z_{MN}\} = Re\{Z_{MN_{opt-\eta}}\}$, the efficiency obtained could be even lower than the one obtained with $Re\{Z_{MN}\} \neq Re\{Z_{MN_{opt-\eta}}\}$. However, if the multi-tapped coil is designed such that $k_{Tx\text{-}Rx}$ and Q_{Rx} are not considerably reduced when changing the ON switch, then the MEP can be achieved with almost no reduction in $\eta_{Link\,max}$.

In [44] this multi-tapped Rx coil is used to track the MPP, while in [45] it was used to track the MEP. Although the architecture discussed in [45] is a bit more complex than the topology presented in Fig. 6.16, it is based on the same idea.

This approach is able to achieve load matching for a wide load range in systems that use a Rx coil with many turns. However, the higher the frequency, the less turns the Rx coil has. In [45] the carrier frequency is 100 kHz, while in [44] it is 6.78 MHz. However, this approach is not feasible in links such as the one analyzed in [25] that operates at 144 MHz, and the optimum Rx coil has only two turns. Additionally, the switch ON-resistance is in series with R_{Rx}, reducing the coil quality factor. At high frequency, where R_{Rx} tends to be very low (as the optimum coil has fewer turns), the ON-resistance of the switch becomes comparable to R_{Rx}, strongly reducing the Rx coil quality factor.

6.5 Combining Adaptive and Nonadaptive Approaches to Achieve the OOP

In this chapter, different adaptive circuits used to track the OOP were presented. To complete this analysis, we would like to highlight that an adaptive OOP tracking method could be used jointly with one or more nonadaptive methods, drawing on the advantages of each approach.

As shown in Sect. 5.6, large resistance conversion ratios can be achieved with matching networks (see Table 5.4). However, as seen in Sect. 6.4.2, the implementation of adjustable matching networks to track the OOP is complex. On the other hand, DC-DC converters can be dynamically adjusted to track the OOP, e.g., by modifying the gain of a switched-capacitor DC-DC converter in Sect. 6.2.2. However, if a large impedance conversion is required, the required gain in the DC-DC converter would be too large or too low, compromising its efficiency.

Therefore, it is advantageous to use a fixed (non-adjustable) matching network for a coarse adjustment and a DC-DC converter (Sect. 6.2) or active rectifier

(Sect. 6.3) for fine-tuning and OOP tracking. In [46], a series-parallel Rx matching network (Fig. 5.13) is used jointly with a reconfigurable DC-DC converter, drawing on the advantages of each approach to achieve and track the OOP.

References

1. M.K. Kazimierczuk, *Pulse-Width Modulated DC-DC Power Converters* (Wiley, 2015)
2. X. Dai, X. Li, Y. Li, A.P. Hu, Maximum efficiency tracking for wireless power transfer systems with dynamic coupling coefficient estimation. IEEE Trans. Power Electron. **33**(6), 5005–5015 (2017)
3. M. Fu, H. Yin, X. Zhu, C. Ma, Analysis and tracking of optimal load in wireless power transfer systems. IEEE Trans. Power Electron. **30**(7), 3952–3963 (2015)
4. Y. Huang, N. Shinohara, T. Mitani, Impedance matching in wireless power transfer. IEEE Trans. Microw. Theory Techn. **65**(2), 582–590 (2017). ISSN 0018-9480. https://doi.org/10.1109/TMTT.2016.2618921
5. M.D. Seeman, S.R. Sanders, Analysis and optimization of switched-capacitor DC-DC converters. IEEE Trans. Power Electron. **23**(2), 841–851 (2008). ISSN 0885-8993
6. H. Lee, Z. Hua, X. Zhang, A reconfigurable 2 × /2.5 × /3 × /4× SC DC-DC regulator with fixed on-time control for transcutaneous power transmission. IEEE Trans. VLSI Syst. **23**(4), 712–722 (2014)
7. P.K. Peter, V. Agarwal, On the input resistance of a reconfigurable switched capacitor DC-DC converter-based maximum power point tracker of a photovoltaic source. IEEE Trans. Power Electron. **27**(12), 4880–4893 (2012)
8. X. Zhang, H. Lee, An efficiency-enhanced auto-reconfigurable 2×/3×SC charge pump for transcutaneous power transmission. IEEE J. Solid State Circuits **45**(9), 1906–1922 (2010). ISSN 0018-9200. https://doi.org/10.1109/JSSC.2010.2055370
9. H.-M. Lee, M. Ghovanloo, An integrated power-efficient active rectifier with offset-controlled high speed comparators for inductively powered applications. IEEE Trans. Circuits Syst. I **58**(8), 1749–1760 (2011)
10. H.M. Lee, M. Ghovanloo, A high frequency active voltage doubler in standard CMOS using offset-controlled comparators for inductive power transmission. IEEE Trans. Biomed. Circuits Syst. **7**(3), 213–224 (2013)
11. L. Cheng, W.H. Ki, Y. Lu, T.S. Yim, Adaptive on/off delay-compensated active rectifiers for wireless power transfer systems. IEEE J. Solid-State Circuits **51**(3), 712–723 (2016). ISSN 0018-9200. https://doi.org/10.1109/JSSC.2016.2517119
12. C. Huang, T. Kawajiri, H. Ishikuro, A near-optimum 13.56 MHz CMOS active rectifier with circuit-delay real-time calibrations for high-current biomedical implants. IEEE J. Solid-State Circuits **51**(8), 1797–1809 (2016). ISSN 0018-9200. https://doi.org/10.1109/JSSC.2016.2582871
13. J. Fuh, S.K. Hsieh, F.B. Yang, P.H. Chen, A 13.56MHz power-efficient active rectifier with digital offset compensation for implantable medical devices, in *IEEE Wireless Power Transfer Conference*, pp. 1–3 (2016). https://doi.org/10.1109/WPT.2016.7498846
14. Z. Xue, D. Li, W. Gou, L. Zhang, S. Fan, L. Geng, A delay time controlled active rectifier with 95.3% peak efficiency for wireless power transmission systems, in *IEEE International Symposium on Circuits and Systems*, pp. 1–4 (2017). https://doi.org/10.1109/ISCAS.2017.8050846
15. P. Pérez-Nicoli, F. Silveira, Reconfigurable multiple-gain active-rectifier for maximum efficiency point tracking in WPT, in *IEEE Latin American Symposium on Circuits and Systems*, pp. 1–4 (2017). https://doi.org/10.1109/LASCAS.2017.7948080

16. R. Mai, Y. Liu, Y. Li, P. Yue, G. Cao, Z. He, An active-rectifier-based maximum efficiency tracking method using an additional measurement coil for wireless power transfer. IEEE Trans. Power Electron. **33**(1), 716–728 (2017)
17. L. Cheng, W.-H. Ki, C.-Y. Tsui, A 6.78-MHz single-stage wireless power receiver using a 3-mode reconfigurable resonant regulating rectifier. IEEE J. Solid-State Circuits **52**(5), 1412–1423 (2017)
18. Z. Li, K. Song, J. Jiang, C. Zhu, Constant current charging and maximum efficiency tracking control scheme for supercapacitor wireless charging. IEEE Trans. Power Electron. **33**(10), 9088–9100 (2018)
19. Z. Huang, S.-C. Wong, K.T. Chi, An inductive-power-transfer converter with high efficiency throughout battery-charging process. IEEE Trans. Power Electron. **34**(10), 10245–10255 (2019)
20. K. Song, R. Wei, G. Yang, H. Zhang, Z. Li, X. Huang, J. Jiang, C. Zhu, Z. Du, Constant current charging and maximum system efficiency tracking for wireless charging systems employing dual-side control. IEEE Trans. Ind. Appl. **56**(1), 622–634 (2019)
21. K. Colak, E. Asa, M. Bojarski, D. Czarkowski, O.C. Onar, A novel phase-shift control of semibridgeless active rectifier for wireless power transfer. IEEE Trans. Power Electron. **30**(11), 6288–6297 (2015)
22. S. Guo, H. Lee, An efficiency-enhanced CMOS rectifier with unbalanced-biased comparators for transcutaneous-powered high-current implants. IEEE J. Solid-State Circuits **44**(6), 1796–1804 (2009)
23. Y. Lu, W.-H. Ki, A 13.56 MHz CMOS active rectifier with switched-offset and compensated biasing for biomedical wireless power transfer systems. IEEE Trans. Biomed. Circuits Syst. **8**(3), 334–344 (2013)
24. G. Bawa, M. Ghovanloo, Analysis, design, and implementation of a high-efficiency full-wave rectifier in standard CMOS technology. Analog. Integr. Circ. Sig. Process. **60**(1), 71–81 (2009). ISSN 1573-1979. https://doi.org/10.1007/s10470-008-9204-7
25. C. Kim, S. Ha, J. Park, A. Akinin, P.P. Mercier, G. Cauwenberghs, A 144-MHz fully integrated resonant regulating rectifier with hybrid pulse modulation for mm-sized implants. IEEE J. Solid-State Circuits **52**(11), 3043–3055 (2017)
26. Q. Li, J. Wang, Y. Inoue, A high efficiency CMOS rectifier with ON-OFF response compensation for wireless power transfer in biomedical applications, in *IEEE International Symposium on Integrated Circuits* (IEEE, 2014), pp. 91–94
27. S. Chen, H. Li, Y. Tang, A burst mode pulse density modulation scheme for inductive power transfer systems without communication modules, in *IEEE Annual Applied Power Electronics Conference and Exposition* (IEEE, 2018), pp. 1071–1075
28. H. Li, J. Fang, S. Chen, K. Wang, Y. Tang, Pulse density modulation for maximum efficiency point tracking of wireless power transfer systems. IEEE Trans. Power Electron. **33**(6), 5492–5501 (2017)
29. U. Guler, M.S.E. Sendi, M. Ghovanloo, A dual-mode passive rectifier for wide-range input power flow, in *IEEE International Midwest Symposium on Circuits and Systems*, pp. 1376–1379 (2017)
30. U. Guler, Y. Jia, M. Ghovanloo, A reconfigurable passive RF-to-DC converter for wireless IoT applications. IEEE Trans. Circuits Syst. II **66**(11), 1800–1804 (2019)
31. U. Guler, Y. Jia, M. Ghovanloo, A reconfigurable passive voltage multiplier for wireless mobile IoT applications. IEEE Trans. Circuits Syst. II **67**(4), 615–619 (2020)
32. H. Lee, An auto-reconfigurable $2 \times /4\times$ AC-DC regulator for wirelessly powered biomedical implants with 28% link efficiency enhancement. IEEE Trans. VLSI Syst. **24**(4), 1598–1602 (2016). ISSN 1063-8210. https://doi.org/10.1109/TVLSI.2015.2452918
33. X. Li, X. Meng, C.-Y. Tsui, W.-H. Ki, Reconfigurable resonant regulating rectifier with primary equalization for extended coupling-and loading-range in bio-implant wireless power transfer. IEEE Trans. Biomed. Circuits Syst. **9**(6), 875–884 (2015)

34. Y. Lu, X. Li, W.-H. Ki, C.-Y. Tsui, C.P. Yue, A 13.56 MHz fully integrated 1X/2X active rectifier with compensated bias current for inductively powered devices, in *IEEE International Solid-State Circuits Conference* (IEEE, 2013), pp. 66–67

35. X. Li, C.Y. Tsui, W.H. Ki, A 13.56 MHz wireless power transfer system with reconfigurable resonant regulating rectifier and wireless power control for implantable medical devices. IEEE J. Solid-State Circuits **50**(4), 978–989 (2015)

36. M. Kiani, B. Lee, P. Yeon, M. Ghovanloo, A Q-modulation technique for efficient inductive power transmission. IEEE J. Solid-State Circuits **50**(12), 2839–2848 (2015). ISSN 0018-9200. https://doi.org/10.1109/JSSC.2015.2453201

37. B. Lee, P. Yeon, M. Ghovanloo, A multicycle Q-modulation for dynamic optimization of inductive links. IEEE Trans. Ind. Electron. **63**(8), 5091–5100 (2016)

38. H.S. Gougheri, M. Kiani, Current-based resonant power delivery with multi-cycle switching for extended-range inductive power transmission. IEEE Trans. Circuits Syst. I **63**(9), 1543–1552 (2016)

39. S. Mandal, R. Sarpeshkar, Power-efficient impedance-modulation wireless data links for biomedical implants. IEEE Trans. Biomed. Circuits Syst. **2**(4), 301–315 (2008)

40. R. Gallichan, D.M. Budgett, D. McCormick, 600mW active rectifier with shorting-control for wirelessly powered medical implants, in *IEEE Biomedical Circuits and Systems Conference* (IEEE, 2018), pp. 1–4

41. Y.-K. Jung, B. Lee, Design of adaptive optimal load circuit for maximum wireless power transfer efficiency, in *IEEE Asia-Pacific Microwave Conference* (IEEE, 2013), pp. 1221–1223

42. B. Lee, M. Kiani, M. Ghovanloo, A triple-loop inductive power transmission system for biomedical applications. IEEE Trans. Biomed. Circuits Syst. **10**(1), 138–148 (2015)

43. J. Tian, A.P. Hu, Stabilising the output voltage of wireless power pickup through parallel tuned DC-voltage controlled variable capacitor. Electron. Lett. **52**(9), 758–759 (2016)

44. P.P. Mercier, A.P. Chandrakasan, Rapid wireless capacitor charging using a multi-tapped inductively-coupled secondary coil. IEEE Trans. Circuits Syst. I **60**(9), 2263–2272 (2013)

45. W. Zhong, S.Y. Hui, Reconfigurable wireless power transfer systems with high energy efficiency over wide load range. IEEE Trans. Power Electron. **33**(7), 6379–6390 (2017)

46. P. Pérez-Nicoli, F. Silveira, Maximum efficiency tracking in inductive power transmission using both matching networks and adjustable AC-DC converters. IEEE Trans. Microw. Theory Techn. **66**(7), 3452–3462 (2018)

Chapter 7
Closed-Loop WPT Links

7.1 Output Voltage Regulation

In an actual WPT link, the distance and alignment between coils may dynamically change, modifying the coupling coefficient, k, which affects the link as studied in Chap. 2. For instance, a WPT link that is established between the external Behind-the-Ear (BTE) part of a hearing aid and the cochlear implant may be affected when the user is jogging or combing his/her hair. In a different scenario, the user may not place his/her smartphone exactly at the center of the wireless charging pad. Additionally, the power required by the load could change, e.g., the loading of a retinal implant would be quite different when the patient is in a dark room versus looking at a bright image. Therefore, in order to provide the power that the load requires at each time, P_L, i.e., maintain the required load voltage, V_L, under P_L and coil position variations, a closed-loop control is usually required. Many different approaches have been proposed for this purpose, which can generally be divided into four categories, as shown in Fig. 7.1: Tx-amplitude (A), Rx matching (B), carrier frequency (C), and Rx-regulator (D), which are described next. In feedback Tx-amplitude (A), V_L is regulated by adjusting the Tx voltage, i.e., adjusting the Tx DC-DC gain, $G_{TxDC\text{-}DC}$, [1–4]. In Rx matching regulation (B), if V_L is higher than the designated value, for instance, when the coils are moved closer to one another and k has increased, the Rx compensation is modified to detune the Rx, reducing the received power in order to regulate V_L [5]. Alternatively, if changes in the environment have resulted in detuning the Rx, an auto-tuning mechanism has been proposed to increase the received power by adjusting the matching circuitry [6]. In carrier frequency regulation (C), the carrier frequency, f, is used as it tunes/detunes not only the Rx, like the Rx-matching regulation (B), but also the Tx [7]. Finally, in feedback Rx-regulator (D), a voltage regulator placed just before the load keeps V_L constant [8]. As discussed in Sect. 6.3, the rectifier can also be in charge of the output voltage regulation, thus implementing the (D)-type feedback and avoiding the use of the DC-DC converter [9, 10].

© The Author(s), under exclusive license to Springer Nature Switzerland AG 2021
P. Pérez-Nicoli et al., *Inductive Links for Wireless Power Transfer*,
https://doi.org/10.1007/978-3-030-65477-1_7

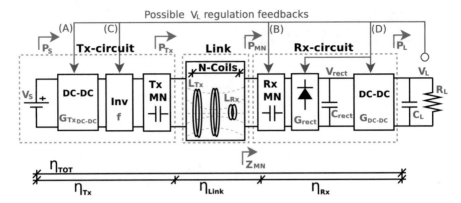

Fig. 7.1 Block diagram of an inductive WPT system, similar to Fig. 1.3, indicating the possible feedback loops to regulate the output voltage, V_L. The $G_{TxDC\text{-}DC}$, $G_{DC\text{-}DC}$, and G_{rect} = (*DC output voltage/peak input voltage*) are the Tx DC-DC, Rx DC-DC, and rectifier gains, respectively

A key difference between feedbacks (A) and (C) vs. (B) and (D) is that the former require back telemetry, while the latter feedbacks do not. On the other hand, (A)-type and (C)-type feedbacks regulate the Tx output power directly and prevent wasting energy and degrading WPT when k increases above its nominal value, while (B)-type and (D)-type feedbacks only make adjustments on the Rx side, neglecting the Tx output power. Therefore, if only a (B)-type or (D)-type feedback is implemented in the WPT link, the Tx should be set to constantly deliver an output power level for the worst-case Tx-Rx coil arrangement scenario to ensure reliable Rx operation in all possible conditions, which could be inefficient. The combination of different feedback control loops, to regulate V_L and maximize the link efficiency, η_{Link}, is discussed in Sect. 7.3.

7.2 Tracking the Maximum Efficiency Point (MEP) in a Closed-Loop

As discussed in Chap. 5, there are optimum values for the Rx coil load impedance, Z_{MN}, which maximize η_{Link} or P_{MN}, achieving the Maximum Efficiency Point (MEP) and Maximum Power Point (MPP), respectively, under a particular condition. In Sect. 5.4, it was proved that the MEP would be the first choice, as high efficiency is always desirable, and thus the system should operate at the MEP as long as the Tx-circuit is capable of delivering the required power in those conditions. Therefore, in this chapter, we will focus on tracking the MEP, in systems where the Tx-circuit is capable of delivering the desired power. In Chap. 5, it was also proved that $Re\{Z_{MN}\}$ depends on the load power consumption, while its optimum value that achieves the MEP, $Re\{Z_{MN_{opt\text{-}\eta}}\}$, depends on the distance and

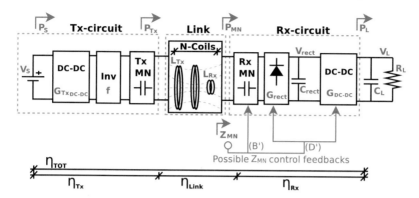

Fig. 7.2 Block diagram of an inductive WPT system, similar to Fig. 1.3, indicating the possible feedback loops to track the MEP. The $G_{TxDC\text{-}DC}$, $G_{DC\text{-}DC}$, and $G_{rect} = (DC\ output\ voltage/peak\ input\ voltage)$ are the Tx DC-DC, Rx DC-DC, and rectifier gains, respectively

alignment between coils. Because of that, different methods have been proposed to automatically track the MEP, $Re\{Z_{MN}\} = Re\{Z_{MN_{opt-\eta}}\}$, against variations in coupling, k, and power consumption, R_L [1, 3, 11, 12]. Some of these methods are graphically represented in Fig. 7.2 and summarized next. The apostrophe in the names of the feedbacks, (B') and (D'), is used to differentiate these feedbacks aimed to track the MEP from the feedbacks aimed to regulate V_L presented in Sect. 7.1. Papers like [11] optimize the Rx compensation through feedback (B'), while [1, 3] and [12] change the gain of the rectifier, G_{rect}, and Rx DC-DC converter, $G_{DC\text{-}DC}$, respectively, through feedback (D'). The use of rectifiers and DC-DC converters in the Rx-circuit to achieve the MEP was previously addressed in Chap. 6. It should be mentioned that other approaches to optimize Z_{MN} exist, such as the Q-modulation addressed in Sect. 6.4.1 [13–15], that are not depicted in Fig. 7.2.

In any actual system, some kind of voltage regulation feedback is required to meet the needs of the load. Therefore, the MEP tracking methods must always operate jointly with a V_L regulation loop. The impact of this on the design of these control loops is further studied in the remaining of the chapter.

7.3   The Joint Use of Output Voltage Regulation and MEP Tracking Feedbacks

All the feedback loops previously introduced, the ones to regulate V_L, presented in Sect. 7.1, and the ones to track the MEP, presented in Sect. 7.2, are represented in Fig. 7.3. The Tx-circuit was substituted by its equivalent Thévenin model to simplify the discussion, without significant loss of generality. X_S is the Tx resulting

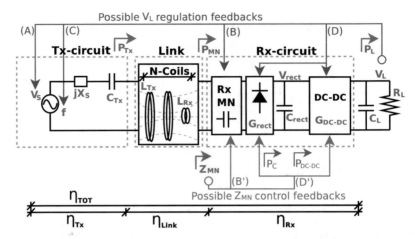

Fig. 7.3 Block diagram of an inductive WPT system indicating the possible feedback loops to regulate the output voltage, V_L, and to track the MEP. In this figure, in contrast to Figs. 7.1 and 7.2, the Tx-circuit was substituted by its equivalent Thévenin model for the sake of simplicity

imaginary part to consider the non-resonant Tx case, which is further addressed in Sect. 7.5.2.

When jointly tracking MEP while regulating V_L, it must be dealt with the fact that in some cases both loops might be trying to alter the same variables with different objectives. For example, any change in the Rx-circuit for regulating V_L like (B) and (D) also changes Z_{MN}, which should be equal to $Z_{MN_{opt-\eta}}$ in order to achieve the MEP. In addition, changes in the carrier frequency, f, (C) also affect the Rx resonance, modifying Z_{MN} and thus the link PTE, η_{Link}.

The straightforward way to achieve both MEP and V_L regulation is to adjust the Rx-circuit in order to hold $Z_{MN_{opt-\eta}}$ and then control the driver based on the value of V_L to regulate it [1, 16]. If, for instance, the power consumption of the load, P_L, changes, a feedback can adapt the Rx-circuit gain, e.g., rectifier gain, G_{rect}, or DC-DC converter gain, G_{DC-DC}, to maintain $Z_{MN} = Z_{MN_{opt-\eta}}$, as discussed in Chap. 6. This change in the gain and P_L will affect V_L, so another feedback has to control it by adjusting the driver, V_S. Therefore, this method is composed of two feedbacks: (A) to control V_L and (B') or (D') to hold $Z_{MN_{opt-\eta}}$. This approach is referred to as preregulation in Fig. 7.4.

A not so intuitive way to achieve the same optimum situation was proposed in [17, 18]. In this approach, the output voltage is regulated by a (D)-type feedback, and $Z_{MN_{opt-\eta}}$ is achieved by adjusting the Tx voltage, V_S, as explained below. This method is referred to as postregulation in Fig. 7.4 and has been used in [8, 18, 19]. It should be noted that changes in the Tx-circuit were not considered to achieve $Z_{MN} = Z_{MN_{opt-\eta}}$ in Figs. 7.2 and 7.3, because actually if only V_S is changed, neither Z_{MN} nor $Z_{MN_{opt-\eta}}$ are altered. However, in postregulation, changes in the Tx-circuit result in changes in the Rx-circuit gain. For example, when a (D)-type feedback is used through the DC-DC converter, $G_{DC-DC} = V_L/V_{rect}$ depends on V_S, because the

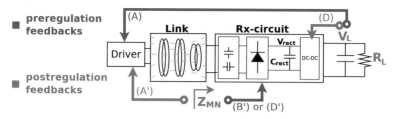

Fig. 7.4 Representation of preregulation and postregulation methods

Table 7.1 Preregulation and postregulation comparison

Preregulation	Postregulation
V_L regulation:	V_L regulation:
• Depends on back telemetry	• No back telemetry needed
MEP tracking:	MEP tracking:
• Z_{MN} is directly controlled in the Rx. $Z_{MN} = Z_{MN_{opt\text{-}\eta}}$ can always be achieved.	• Z_{MN} is indirectly controlled $Z_{MN} = Z_{MN_{opt\text{-}\eta}}$ may not be possible to achieve

rectifier output voltage, V_{rect}, is affected by V_S while V_L is kept constant by the (D)-type feedback. Therefore, the changes in V_S are actually generating changes in the Rx-circuit modifying Z_{MN}, and thus it could be used to achieve $Z_{MN} = Z_{MN_{opt\text{-}\eta}}$. This is referred to as feedback (A') in Fig. 7.4. The dynamic of postregulation is more complex to analyze, and it is not easy to determine at first glance if the MEP, $Z_{MN} = Z_{MN_{opt\text{-}\eta}}$, can be achieved. Indeed, in [19], it was pointed out that in the particular case, where a voltage source is used to drive a resonant Tx in a 2-coil link with a series Rx compensation, the MEP cannot be achieved if only V_S is adjusted. Additionally, it was proved that η_{Link} is limited in that case to 50% even for ideal coils (Q_{Tx} and $Q_{Rx} \to \infty$) where, theoretically, η_{Link} should tend to 100% [20].

Table 7.1 presents a brief comparison between preregulation and postregulation methods. Since postregulation controls V_L directly in the Rx, it is more stable under k and R_L variations as long as sufficient power is delivered through the link. In preregulated links [1, 2], a Low-Dropout Regulator (LDO) is usually added to improve V_L stability. However, this LDO increases losses, especially if its input voltage, $V_{LDO_{in}}$, is much larger than its output voltage, V_L. If, in order to reduce the LDO losses, the LDO input voltage is adjusted to $V_{LDO_{in}} \cong V_L$, a sudden decrease in the coupling between coils may turn OFF the LDO until the feedback amends the Tx-circuit. This adjustment of the Tx-circuit takes time as it depends on the back telemetry throughput and the inductive link response time.

In preregulation, Z_{MN} is directly controlled in the Rx which assures that $Z_{MN_{opt\text{-}\eta}}$ is reached, as studied in Chap. 6. In postregulation, $Z_{MN_{opt\text{-}\eta}}$ can be tracked from the driver without any back telemetry, as explained below. Assuming that the Rx-circuit is working correctly and able to deliver the P_L demanded by the load, the MEP is tracked by adjusting V_S until the Tx output power, P_{Tx}, is minimized which can be measured directly in the Tx [17, 18]. This minimum P_{Tx} is usually reached by a Perturbation and Observation (P&O) method. When P_{Tx} is minimized, $Z_{MN_{opt-\eta}}$ is indirectly achieved. However, this is not possible in all postregulated systems as will be discussed in Sect. 7.5.

7.4   Tracking the MEP in Links with Preregulated Output Voltage

As was depicted in Fig. 7.4, a preregulated link is composed of an (A) feedback, which regulates V_L, and a (B') or (D') feedback, which controls Z_{MN} to track the MEP.

Although feedback (A) requires back telemetry, its implementation is simple. It just requires to know the value of V_L in the Tx and adjust the Tx output power in closed-loop to regulate V_L.

Regarding the implementation of feedback (B') or (D'), although any value of Z_{MN} can be set by using a multiple gain rectifier or DC-DC converter in the Rx, as studied in Chap. 6, the MEP tracking in closed-loop is not trivial. This is because the desired target value, $Z_{MN_{opt-\eta}}$, depends on the coupling coefficient, k, which could variate during link operation. The coupling coefficient, k, could be dynamically measured, to calculate $Z_{MN_{opt-\eta}}$ and adjust Z_{MN} accordingly. However, dynamically measuring k, as addressed in [21], is complex. A simpler way to implement feedbacks (B') or (D') is using a P&O method, as proposed in [12]. In [12], the duty cycle of the cascade of a boost and a buck converter in the Rx is modified (perturbation); then the efficiency is calculated (observation), by measuring the input and output powers; and the new value of efficiency is compared with the one obtained in the previous step, in order to track the MEP. The load voltage, V_L, and load power, P_L, are adjusted by the (A) feedback as in [22].

Summarizing, if the Rx-circuit is well designed and able to correctly variate Z_{MN}, as analyzed in Chap. 6, a P&O method will always achieve the MEP by the (B') or (D') feedbacks in a preregulated link. However, as is going to be studied in the next section, in postregulated links, the (A') feedback, which was depicted in Fig. 7.4, is not always able to achieve the MEP. Therefore, in order to ensure the MEP attainability in postregulated links, a further analysis is required, which is presented next.

7.5   Tracking the MEP in Links with Postregulated Output Voltage

The postregulated links deserve special analysis as the MEP cannot always be achieved, which is addressed in this section.

Let us consider the postregulated system of Fig. 7.5, where a (D)-type feedback is implemented through the DC-DC converter to regulate V_L and the MEP is achieved through an (A')-type feedback.

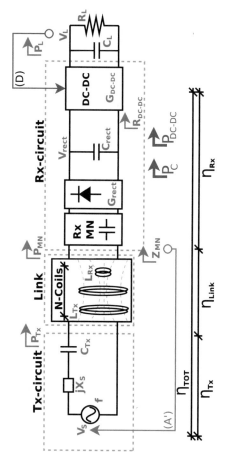

Fig. 7.5 Postregulated system

The purpose of this section is to study how the MEP, $Z_{MN} = Z_{MN_{opt-\eta}}$, can be achieved in postregulated links, which requires to further analyze the dependency between the Rx-circuit input impedance, Z_{MN}, and V_S, i.e., the feedback (A'). Therefore, in this section, we first address how the value of Z_{MN} is determined in a postregulated system and the range of values it can take when V_S is modified. Additionally, we will analyze if the range of values of Z_{MN} under V_S variations includes the optimum desired value, $Z_{MN_{opt-\eta}}$, which determines whether the MEP can be achieved. So let us start with the following analysis that will allow us to deduce the relationship between Z_{MN} and V_S.

In steady-state condition, the output voltage of the rectifier, V_{rect}, is constant, and the rectifier output power, P_C, is equal to the DC-DC regulator input power, P_{DC-DC}, because C_{rect} is an ideal capacitor. However, we keep these two parameters separate for the purpose of our discussion.

Regardless of the selected approach, if V_L is kept constant, P_L and thus P_{DC-DC} (approximating η_{DC-DC} as a constant) are determined by the load and independent of the previous blocks in the system. However, the rectifier output power, P_C, basically depends on all the parameters of the system. This power, P_C, is affected by the following: first, by the Tx-circuit, V_S, X_S, and f; second by the link, number of coils, N, coil quality factor Q_i where $i \in [1, N]$, and coupling coefficient (k_{ij}) where $i, j \in [1, N] i \neq j$; and third by the Rx matching network, η_{MN}, and rectifier, η_{rect}, efficiencies. Finally, P_C also depends on the DC-DC converter input resistance R_{DC-DC}, which affects the Rx coil load impedance, Z_{MN}, as studied in Chap. 6. This resistance can be estimated as

$$P_{DC-DC} = \frac{V_{rect}^2}{R_{DC-DC}} \Rightarrow R_{DC-DC} = V_{rect}^2 / P_{DC-DC}, \qquad (7.1)$$

thus it depends on V_{rect}. The steady-state condition, $P_C = P_{DC-DC}$, is presented in (7.2) where all the parameters that affect P_C are listed:

$$P_C = f(V_S, X_S, f, N, Q_i, k_{ij}, \eta_{MN}, \eta_{rect}, V_{rect}) = P_{DC-DC} \qquad (7.2)$$

The designer could directly set or determine all the parameters presented in (7.2) except for V_{rect}. In consequence, for a given set of parameters, V_{rect} should converge to a value that satisfies (7.2).

To better understand the steady-state condition of (7.2) and to write it as a function of the variable of interest, Z_{MN}, let us consider Fig. 7.6 where the Tx-circuit and the inductive link are modeled by its equivalent Thévenin circuit. From Fig. 7.6, the equality of (7.2) can be rewritten as

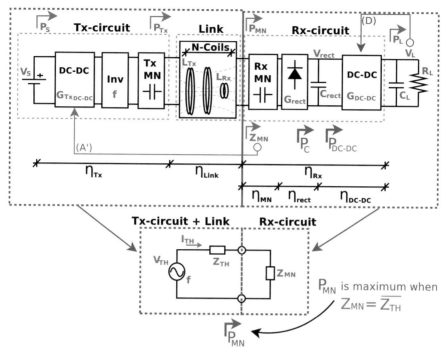

Fig. 7.6 Postregulated system where the Tx-circuit and the inductive link are both modeled by its equivalent Thévenin model

$$\underbrace{\underbrace{f(V_{TH}, Z_{TH}, Z_{MN})}_{P_C} \cdot \eta_{MN} \cdot \eta_{rect}}_{P_{MN}} = \underbrace{\frac{\overbrace{\frac{P_L}{\eta_{DC\text{-}DC}}}^{P_{DC\text{-}DC}}}{}}_{} \Rightarrow$$

$$\underbrace{\underbrace{\frac{\overbrace{V_{TH}^2}^{}}{\underbrace{|Z_{TH} + Z_{MN}|^2}_{I_{TH}^2} \frac{1}{2} Re\{Z_{MN}\}}}_{P_{MN}} \cdot \eta_{MN} \cdot \eta_{rect}}_{} = \frac{\overbrace{\dfrac{P_L}{\eta_{DC\text{-}DC}}}^{P_{DC\text{-}DC}}}{} . \tag{7.3}$$

To simplify the analysis, we will approximate η_{MN}, η_{rect}, and $\eta_{DC\text{-}DC}$ by constants. These approximations are needed in order to analyze the system in general and will be valid in the operating range of these blocks. We will also assume that the Rx-circuit is designed to ensure the Rx is at resonance, meaning $Im\{Z_{MN}\} = -\omega L_{Rx}$, and only its real part, $Re\{Z_{MN}\}$, which depends on V_{rect}, remains. This is a desired behavior that is validated during the analysis.

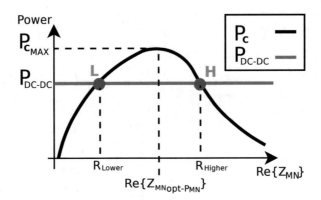

Fig. 7.7 A simplified graph showing the changes in P_C as a function of $Re\{Z_{MN}\}$, used to analyze the operating point and its dependence on the system parameters. The $P_{DC\text{-}DC}$, which is equal to $P_L/\eta_{DC\text{-}DC}$, is also represented. The intersection points, L and H, are the operating points that satisfy the steady-state condition (7.3)

In Fig. 7.7, the left side of equality (7.3), P_C, and the right side of it, $P_{DC\text{-}DC}$, are sketched as a function of $Re\{Z_{MN}\}$.

As explained before, P_L is set by the load, and it does not depend on $Re\{Z_{MN}\}$. Therefore, $P_{DC\text{-}DC} = P_L/\eta_{DC\text{-}DC}$ is constant and represented in Fig. 7.7 by a horizontal line. On the other hand, P_C has a maximum value at $Re\{Z_{MN_{opt\text{-}P_{MN}}}\}$. This behavior corresponds to the well-known maximum power transfer theorem, where the maximum power transfer occurs when Z_{MN} is the conjugate of the Tx-circuit+Link output impedance, $\overline{Z_{TH}}$, as shown in Fig. 7.6. This optimum value of $Re\{Z_{MN}\}$ that maximizes P_{MN} was studied in Chap. 5 where the same Thévenin model was introduced.

As can be seen from Fig. 7.7, if $P_{DC\text{-}DC}$ is too high, $P_{DC\text{-}DC} > P_{C_{MAX}}$, there is no $Re\{Z_{MN}\}$ that satisfies (7.3). This means that the system is not going to work as the power demanded by the load cannot be provided. If $P_{DC\text{-}DC} < P_{C_{MAX}}$, two values of $Re\{Z_{MN}\}$ that satisfy (7.3) exist, R_{Lower}, which is lower than the value that gives maximum P_C power, $Re\{Z_{MN_{opt\text{-}P_{MN}}}\}$, and R_{Higher}, which is higher than $Re\{Z_{MN_{opt\text{-}P_{MN}}}\}$, as shown in Fig. 7.7. One of these values is stable, while the other is unstable depending on the Rx-circuit as will be shown in Sect. 7.5.1. It should also be noted that when $P_{DC\text{-}DC} = P_{C_{MAX}}$, $Re\{Z_{MN}\}$ satisfies (7.3) in only one point.

The sketch of Fig. 7.7 was done for a given value of source Thévenin equivalent voltage, V_{TH}. If V_{TH} scales up, the P_C curve, which is proportional to V_{TH}^2 (7.3), also scales up, as depicted in Fig. 7.8. Therefore, V_{TH} can be dynamically modified by changing the Tx voltage, V_S, to adjust $Re\{Z_{MN}\}$. Depending on the operating point, "lower" (L) or "higher" (H), $Re\{Z_{MN}\}$ ranges from zero to $Re\{Z_{MN_{opt\text{-}P_{MN}}}\}$ or from $Re\{Z_{MN_{opt\text{-}P_{MN}}}\}$ to infinity, respectively, as depicted in Fig. 7.8. Therefore, as pointed out before, V_S can be used to modify $Re\{Z_{MN}\}$ and try to achieve the MEP. This was referred to as feedback (A') in Fig. 7.5. The MEP can be achieved

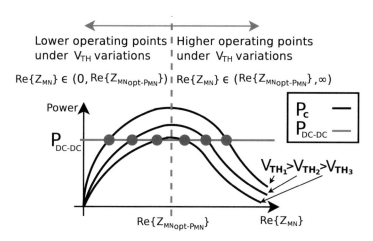

Fig. 7.8 Ranges of values of $Re\{Z_{MN}\}$ when sweeping V_{TH}

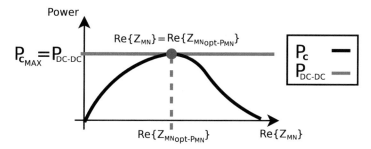

Fig. 7.9 Particular situation where V_S is such that $P_{DC\text{-}DC}$ is equal to the maximum P_C, $P_{DC\text{-}DC} = P_{C_{MAX}}$, thus $Re\{Z_{MN}\} = Re\{Z_{MN_{opt\text{-}P_{MN}}}\}$ achieving the MPP

only if the $Re\{Z_{MN}\}$ that maximizes η_{Link}, $Re\{Z_{MN_{opt\text{-}\eta}}\}$, belongs to the achievable interval of $Re\{Z_{MN}\}$.

The MPP, $Re\{Z_{MN}\} = Re\{Z_{MN_{opt\text{-}P_{MN}}}\}$, can be achieved only if V_S is such that $P_{DC\text{-}DC}$ is equal to the maximum P_C, $P_{DC\text{-}DC} = P_{C_{MAX}}$; this particular situation is represented in Fig. 7.9.

The previous analysis helps us to understand how Z_{MN} is determined in a postregulated system and illustrates how V_S can be used to modify its value, i.e., the (A') feedback loop. In the remainder of the chapter, the attainability of the MEP in postregulated links is studied. First, in Sect. 7.5.1, the Rx-circuit is studied to prove how it determines the stability and instability of points L and H of Figs. 7.7 and 7.8. Next, the MEP attainability is addressed in 2-coil and 3-coil links in Sects. 7.5.2 and 7.5.3, respectively. Finally, a generalization to an N-coil link is presented in Sect. 7.5.4 followed by measurement results of 2-, 3-, and 4- coil links in Sect. 7.5.5.

7.5.1   Effect of Rx-circuit in the Operating Point

In this section, the stability of the L and H operating points in Fig. 7.7 is studied.
To do this, we analyze the system behavior before reaching the L or H points. First,
let us consider the situation represented in Fig. 7.10, where the system is receiving
more power than what is required by the load. In that situation, the rectifier output
power, P_C, is greater than the DC-DC converter input power, $P_{DC\text{-}DC}$; thus, the
energy stored in C_{rect} will start to increase, increasing V_{rect}. Since the input power
of the DC-DC converter is approximately constant, the larger V_{rect} is, the lower its
input current is, i.e., its input resistance, $R_{DC\text{-}DC}$, increases with V_{rect} as shown in
(7.1). This modification in $R_{DC\text{-}DC}$ affects $Re\{Z_{MN}\}$, which is the horizontal axis in
Fig. 7.10. Therefore, in order to determine the stable operating point in steady state,
we should deduce if the change in $R_{DC\text{-}DC}$ will result in the operating point moving
to the left, by reducing $Re\{Z_{MN}\}$, or to the right, by increasing $Re\{Z_{MN}\}$.

The relationship between $R_{DC\text{-}DC}$ and Z_{MN} depends on the Rx matching network
used. In Table 5.4, the input impedance of different matching networks was studied.
In Table 7.2, the impedance conversion for the typical series and parallel matching
networks are recalled, together with an approximation of the rectifier impedance
conversion. Using the equations presented in Table 7.2, the relationship between
$R_{DC\text{-}DC}$ and Z_{MN} can be found. Regarding the rectifier, note that (6.3), deduced in
Sect. 6.3.2, neglects the harmonic distortion that it generates. We only use (6.3) to
conclude that the rectifier input resistance, R_{rect}, increases with its load resistance,
$R_{DC\text{-}DC}$. Then, depending on the Rx matching network, this increase in R_{rect} will
increase $Re\{Z_{MN}\}$ in the series case (7.4) or decrease $Re\{Z_{MN}\}$ in the parallel
case (7.5). The previous discussion regarding the situation presented in Fig. 7.10
is summarized in (7.6) and (7.7). In conclusion, with a series matching network,
the Rx-circuit tends to increase the $Re\{Z_{MN}\}$ when it is receiving more power than
required. Contrary to the series case, the parallel matching network tends to decrease
the $Re\{Z_{MN}\}$ in that situation.

The same behavior deduced for the series matching network occurs for any
matching network with a direct relationship between R_{rect} and $Re\{Z_{MN}\}$. Analogously, any matching network with an inverse relationship between R_{rect} and
$Re\{Z_{MN}\}$ behaves as the parallel case.

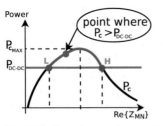

Fig. 7.10 Operating point analysis.
Case where P_C is greater than $P_{DC\text{-}DC}$

$$P_C > P_{DC\text{-}DC} \Rightarrow \begin{array}{c} \text{increases the energy} \\ \text{stored in } C_{rect} \end{array} \Rightarrow \tag{7.6}$$

$$V_{rect} \uparrow \xrightarrow{(7.1)} R_{DC\text{-}DC} \uparrow \xrightarrow{(6.3)} R_{rect} \uparrow \Rightarrow$$

- **Series** $\xrightarrow{(7.4)} Re\{Z_{MN}\} \uparrow$

- **Parallel** $\xrightarrow{(7.5)} Re\{Z_{MN}\} \downarrow$ $\qquad$ (7.7)

Table 7.2 Impedance transformation

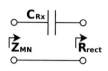	Series Rx matching network: $$Z_{MN} = \overbrace{R_{rect}}^{Re\{Z_{MN}\}} + \frac{1}{j\omega C_{Rx}} \qquad (7.4)$$
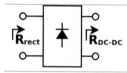	Parallel Rx matching network: (if $R_{rect}^2/(\omega L_{Rx})^2 \gg 1$, $\omega L_{Rx} = 1/\omega C_{Rx}$) $$Z_{MN} = \overbrace{\frac{(\omega L_{Rx})^2}{R_{rect}}}^{Re\{Z_{MN}\}} + \frac{1}{j\omega C_{Rx}} \qquad (7.5)$$
	Rectifier: $$R_{rect} \simeq \frac{R_{DC\text{-}DC} \cdot \eta_{rect}}{2G_{rect}^2} \qquad (6.3)$$

Note: R_{rect} (6.3) was deduced in Sect. 6.3.2, neglecting parasitic capacitances, the diode threshold voltage, and all the harmonic generation

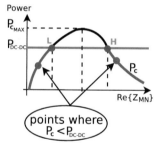

Fig. 7.11 Operating point analysis. Case where P_C is less than $P_{DC\text{-}DC}$

$$P_C < P_{DC\text{-}DC} \Rightarrow \begin{array}{c} \text{decreases the energy} \\ \text{stored in } C_{rect} \end{array} \Rightarrow$$
$$V_{rect} \downarrow \xrightarrow{(7.1)} R_{DC\text{-}DC} \downarrow \xrightarrow{(6.3)} R_{rect} \downarrow \Rightarrow \qquad (7.8)$$

- **Series** $\xrightarrow{(7.4)} Re\{Z_{MN}\} \downarrow$
- **Parallel** $\xrightarrow{(7.5)} Re\{Z_{MN}\} \uparrow$
$\qquad (7.9)$

The case where P_C is less than $P_{DC\text{-}DC}$, represented in Fig. 7.11, can be studied in the same way and using the same equations previously presented. In this case $P_C < P_{DC\text{-}DC}$; thus the energy stored in the capacitor C_{rect} tends to decrease, reducing V_{rect}. As can be seen from Fig. 7.11, (7.8) and (7.9), when $P_C < P_{DC\text{-}DC}$, the Rx-circuit tends to reduce $Re\{Z_{MN}\}$ if a series matching network is used, while it tends to increase $Re\{Z_{MN}\}$ if a parallel matching network is used.

The results of both cases presented in Figs. 7.10 and 7.11, for the series and the parallel matching network, are graphically represented in Fig. 7.12a, b, respectively. The arrowheads indicate the previously discussed dynamic of the system, i.e., if $Re\{Z_{MN}\}$ tends to increase or decrease from a given value. An operating point is unstable when a small perturbation tends to move it further away, which is indicated by the outgoing arrowheads from the operating point. On the other hand,

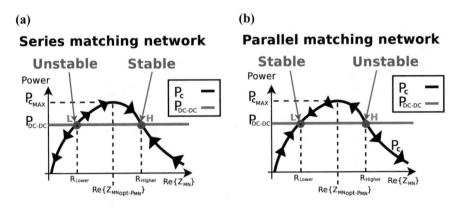

Fig. 7.12 Graphical representation of the analysis for the case of a series and parallel matching network. (**a**) With series matching network (**b**) With parallel matching network

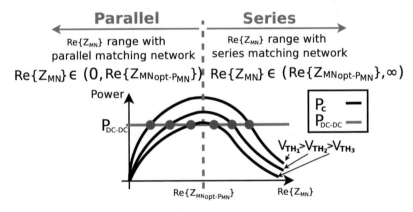

Fig. 7.13 Range of values of $Re\{Z_{MN}\}$ under V_S variations, for the series and the parallel matching network

an operating point is stable when the system tends to amend small perturbation of the operating point, which is indicated by the incoming arrowheads to the operating point. Although this is not a formal stability analysis, it lets us perceive which point will be a stable solution and which will not.

Based on the stability of the "lower" (L) and "higher" (H) points, it can be seen from Fig. 7.12a, b that a system with a series matching network operates in the "higher" (H) point, while a system with a parallel matching network operates in the "lower" (L) point.

Figure 7.8, where the ranges of values of $Re\{Z_{MN}\}$ were illustrated, can be complemented with the outcomes of this section. Now the "lower" and "higher" points correspond to the parallel and series matching network, respectively, as shown in Fig. 7.13.

To know if the MEP, $Re\{Z_{MN_{opt-\eta}}\}$, can be achieved, we still need to know which range, the series (higher) or parallel (lower), contains the $Re\{Z_{MN_{opt-\eta}}\}$. The answer depends on the Tx-circuit and the number of coils used in the inductive link, which is studied in the remainder of the chapter.

7.5.2   2-Coil Links

From the attainable intervals of $Re\{Z_{MN}\}$ represented in Fig. 7.13, it is known that when a series Rx matching network is used, the MEP, $Re\{Z_{MN}\} = Re\{Z_{MN_{opt-\eta}}\}$, can only be achieved if $Re\{Z_{MN_{opt-\eta}}\} > Re\{Z_{MN_{opt-P_{MN}}}\}$. On the contrary, when a parallel Rx matching network is used, the MEP can only be achieved if $Re\{Z_{MN_{opt-\eta}}\} < Re\{Z_{MN_{opt-P_{MN}}}\}$. Therefore, the attainability of the MEP depends not only on the Rx matching network used (series or parallel) but also on whether $Re\{Z_{MN_{opt-\eta}}\}$ is greater than or less than $Re\{Z_{MN_{opt-P_{MN}}}\}$.

For the 2-coil link, the values of $Re\{Z_{MN_{opt-\eta}}\}$ and $Re\{Z_{MN_{opt-P_{MN}}}\}$ were deduced in Sects. 5.2 and Sect. 5.3, respectively. It was shown that the $Re\{Z_{MN}\}$ that maximizes η_{Link} is

$$Re\{Z_{MN_{opt-\eta}}\} = R_{Rx}\sqrt{k_{Tx\text{-}Rx}^2 Q_{Tx} Q_{Rx} + 1}. \tag{5.4}$$

Regarding the $Re\{Z_{MN}\}$ that maximizes P_{MN}, it was proved that its value depends on the Tx-circuit used. With a voltage source and a series resonant Tx

$$Re\{Z_{MN_{opt-P_{MN}}}\} = R_{Rx}(k_{Tx\text{-}Rx}^2 Q_{Tx} Q_{Rx} + 1). \tag{5.9}$$

while with a current source and a series resonant Tx

$$Re\{Z_{MN_{opt-P_{MN}}}\} = R_{Rx}. \tag{5.16}$$

Comparing (5.4), (5.9), and (5.16)

Current source   **Voltage source**

series resonant Tx   **series resonant Tx**

$$\underbrace{Re\{Z_{MN_{opt-P_{MN}}}\}}_{R_{Rx}} < \overbrace{R_{Rx}\sqrt{k_{Tx\text{-}Rx}^2 Q_{Tx} Q_{Rx} + 1}}^{Re\{Z_{MN_{opt-\eta}}\}} < \overbrace{R_{Rx}(k_{Tx\text{-}Rx}^2 Q_{Tx} Q_{Rx} + 1)}^{Re\{Z_{MN_{opt-P_{MN}}}\}} \tag{7.10}$$

The values of $Re\{Z_{MN_{opt-\eta}}\}$ and $Re\{Z_{MN_{opt-P_{MN}}}\}$ are graphically represented in Fig. 7.14.

Therefore, in a postregulated 2-coil link, with a series resonant Tx, the MEP attainability depends on the type of driver (current or voltage) and Rx matching network, as summarized in Table 7.3.

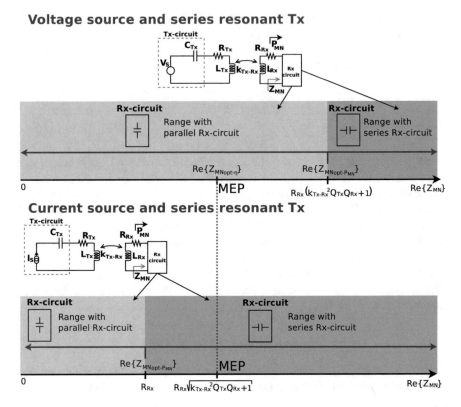

Fig. 7.14 Range of values of $Re\{Z_{MN}\}$ in a 2-coil link, for series Rx matching network, parallel Rx matching network, voltage Tx driver, and current Tx driver

The simulation results of two example systems are presented in Table 7.4, for the cases I and II of Table 7.3. In both cases, we plot $Re\{Z_{MN}\}$ and η_{Link} as a function of V_S (Figs. 7.15 and 7.16). In these examples, V_S was decreased until $V_S \simeq 1$ V because with $V_S < 1$ V the WPT link is not able to deliver the power required by the load, i.e., $P_{C_{MAX}}$ drops below $P_{DC\text{-}DC}$ (see Fig. 7.7). The $Re\{Z_{MN}\}$ was calculated from the first harmonic of the Rx matching network input current, I_{MN}, and input voltage, V_{MN}, as $Re\{Z_{MN}\} = Re\{V_{MN}/I_{MN}\}$. As can be seen, the ranges of values of $Re\{Z_{MN}\}$ match the theoretical ranges represented in Fig. 7.14, from zero to $Re\{Z_{MN_{opt\text{-}P_{MN}}}\} = 25$ Ω and from $Re\{Z_{MN_{opt\text{-}P_{MN}}}\} = 25$ Ω to infinity, for the parallel and series topology, respectively. In this case, $Re\{Z_{MN_{opt\text{-}\eta}}\} = 11$ Ω, which is within the range of values of $Re\{Z_{MN}\}$ that are achieved with the parallel topology. Therefore, it is the parallel matching network which is able to achieve the MEP of $\eta_{Link} = 38\%$, while the series matching network achieves a maximum efficiency of only 31% and it stops working, i.e., receives less power than required by the load, at $V_S \simeq 1$ V, without reaching the MEP.

Table 7.3 Summary of the different cases studied for the 2-coil link, indicating whether MEP tracking is possible by adjusting the Tx-circuit amplitude

Case	Tx-circuit	Rx matching network	Attainability of MEP
I	Voltage source	Series	NO
II	Voltage source	Parallel	YES
III	Current source	Series	YES
IV	Current source	Parallel	NO

Whenever possible, the system should be designed like cases II or III of Table 7.3 to be able to achieve the MEP. The choice between series and parallel Rx matching network affects not only the MEP attainability but also the value of Z_{MN} as calculated in Table 7.2. Although in theory it should be possible to achieve the MEP with any Rx matching network, it may require the Rx DC-DC to be operating with a too high (or low) gain, which would be inefficient. When the matching network is designed in such a way that the $Re\{Z_{MN}\}$ obtained with unitary gain in the Rx DC-DC converter is too far from the $Re\{Z_{MN_{opt-\eta}}\}$, it would be impossible or inefficient to adjust it using the active circuits presented in Chap. 6, i.e., active rectifiers or DC-DC converters, as pointed out in Sect. 6.2.2. Therefore, it is the Tx-circuit architecture which should be selected to be in cases II or III of Table 7.3 in postregulated systems, while the Rx matching network is selected to have the $Re\{Z_{MN}\}$ as close as possible to the optimum value, $Re\{Z_{MN_{opt-\eta}}\}$, with a unitary gain in the Rx DC-DC converter.

An alternative to using a current source in the Tx is presented next, where it is proved that a non-resonant Tx behaves like a current source and could be used to attain the MEP.

7.5.2.1 Analysis with Non-resonant Tx-circuit

So far, in this section, we have considered a series resonant Tx-circuit. Now, the non-resonant Tx case is considered to explore its influence on the MEP attainability.

The circuit schematic of the 2-coil link with a non-resonant Tx is shown in Fig. 7.17. We still assume that the Rx coil is always resonating, $Im\{Z_{MN}\} = -\omega L_{Rx}$, as this maximizes the link efficiency, η_{Link}. However, an extra imaginary part is considered in the Tx, jX_S, as it does not jeopardize the link efficiency but it does affect the operating point as will be shown. To simplify the discussion, in Fig. 7.17, the series resonant capacitor in the Tx is assumed to be cancelling the coil impedance, $j\omega L_{Tx}$, leaving only jX_S as the series reactance. This extra reactance can be generated, in an actual case, by detuning the Tx or using a general matching network as it is used in the Rx. The real part of the output impedance of the driver can be included as part of the coil parasitic resistance, R_{Tx}, to simplify the equations.

Table 7.4 Simulation result for cases I and II of Table 7.3. The LTspice schematics and calculation script are available in the supplementary material, files Sec752, Sec752_series and Sec752_parallel

---

Parameters for the table:
$R_{Tx} = R_{Rx} = 5\ \Omega$; $Q_{Tx} = Q_{Rx} = 200$; $k_{Tx\text{-}Rx} = 0.01$; $f = 13.56$ MHz;
$\eta_{MN} = 100\ \%$; $\eta_{rect} = $ (aprox) 100%; $\eta_{(DC\text{-}DC)} = 80\ \%$; $V_L = 4$ V; $R_L = 1$ kΩ; $P_L = 16$ mW

With this parameters:
$(5.9) \Rightarrow Re\{Z_{MN_{opt\text{-}P_{MN}}}\} = 25\ \Omega$
$(5.4) \Rightarrow Re\{Z_{MN_{opt\text{-}\eta}}\} = 11\ \Omega$
$(5.5) \Rightarrow \eta_{Link\,max} = 38\ \%$

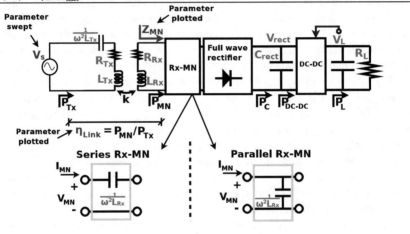

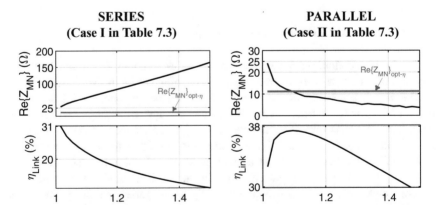

SERIES (Case I in Table 7.3)	PARALLEL (Case II in Table 7.3)

Fig. 7.15 Range of values of $Re\{Z_{MN}\}$ and η_{Link} when V_S is swept in a 2-coil link with series Rx matching network

Fig. 7.16 Range of values of $Re\{Z_{MN}\}$ and η_{Link} when V_S is swept in a 2-coil link with parallel Rx matching network

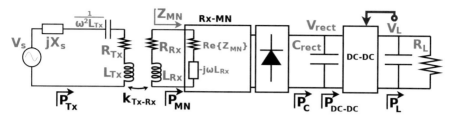

Fig. 7.17 Schematic circuit for a postregulated 2-coil link with non-resonant Tx-circuit

The rectifier output power, P_C, in this non-resonant Tx case, is presented in (7.11), and the value of $Re\{Z_{MN}\}$ that maximizes it, $Re\{Z_{MN_{opt-P_{MN}}}\}$, is presented in (7.12). The deduction of (7.11) and (7.12) can be found in Appendix C.1:

$$P_C = \overbrace{\frac{Q_{Rx\text{-}L}}{Q_L} \frac{k_{Tx\text{-}Rx}^2 Q_{Tx} Q_{Rx\text{-}L}}{(1 + k_{Tx\text{-}Rx}^2 Q_{Tx} Q_{Rx\text{-}L})^2 + \left(\frac{X_S}{R_{Tx}}\right)^2} \frac{V_S^2}{2R_{Tx}}}^{P_{MN}} \eta_{MN} \eta_{rect} \qquad (7.11)$$

$$Re\{Z_{MN_{opt-P_{MN}}}\} = R_{Rx} \sqrt{\frac{(1 + k_{Tx\text{-}Rx}^2 Q_{Tx} Q_{Rx})^2 + (X_S/R_{Tx})^2}{(X_S/R_{Tx})^2 + 1}} \qquad (7.12)$$

Since the non-resonance of the Tx does not affect the theoretical expression of the link efficiency, η_{Link} (2.13), the $Re\{Z_{MN}\}$ that maximizes η_{Link}, $Re\{Z_{MN_{opt-\eta}}\}$, continue being the same as the resonant case previously presented in (5.4), and does not depend on X_S. However, P_C (7.11) and $Re\{Z_{MN_{opt-P_{MN}}}\}$ (7.12) do depend on X_S. Thus, by modifying X_S, $Re\{Z_{MN_{opt-P_{MN}}}\}$ is altered, while $Re\{Z_{MN_{opt-\eta}}\}$ remains unchanged. As previously studied, the MEP attainability depends on whether $Re\{Z_{MN_{opt-P_{MN}}}\}$ is greater (or less) than $Re\{Z_{MN_{opt-\eta}}\}$, which, in turn, depends on X_S.

By modifying X_S, $Re\{Z_{MN_{opt-P_{MN}}}\}$ can range as shown in (7.13) and Fig. 7.18:

$$(7.12) \Rightarrow Re\{Z_{MN_{opt-P_{MN}}}\} \in \left(\underbrace{R_{Rx}}, \underbrace{R_{Rx}(k_{Tx\text{-}Rx}^2 Q_{Tx} Q_{Rx} + 1)} \right] \qquad (7.13)$$

$$\lim_{|X_S| \to \infty} Re\{Z_{MN_{opt-P_{MN}}}\}$$

$$\lim_{|X_S| \to 0} Re\{Z_{MN_{opt-P_{MN}}}\}$$

The left and the right edges of the $Re\{Z_{MN_{opt-P_{MN}}}\}$ interval (7.13) match with the $Re\{Z_{MN_{opt-P_{MN}}}\}$ of a series resonant current source (5.16) and voltage source (5.9), respectively. This is because when $|X_S| \to 0$, the system becomes resonant, and when $|X_S| \to \infty$, the Tx-circuit behaves as a current source as its output current is almost not affected by its load impedance which is negligible compared to $|X_S| \to \infty$.

Even without reaching the edges of the interval of (7.13), it is possible to set $Re\{Z_{MN_{opt-P_{MN}}}\}$ to be greater or less than $Re\{Z_{MN_{opt-\eta}}\}$ by modifying X_S, as shown in Fig. 7.18. This means that the case with a voltage driver and a series Rx matching network, which is unable to track the MEP if the Tx is resonating, case I in Table 7.3, could reach the MEP if the Tx is detuned enough. This detune in the Tx-circuit to achieve the MEP is graphically represented in Fig. 7.19.

Increasing $|X_S|$ allows us to reach the MEP, despite some side effects. First, as can be seen from (7.11), the higher $|X_S|$ is, a higher V_S is required in order to deliver the same amount of power. Second, the operation of the driver circuit could be affected if it is loaded with a reactive loading. This would depend on the architecture used for the driver circuit and should be considered in the system design.

Non-resonant voltage source and series resonant Tx

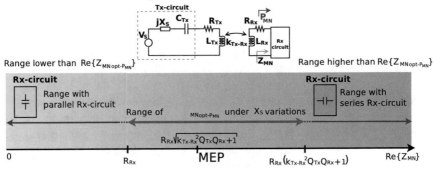

Fig. 7.18 Range of values of $Re\{Z_{MN_{opt-P_{MN}}}\}$ under X_S variations

Non-resonant voltage source and series resonant Tx

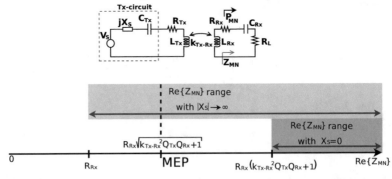

Fig. 7.19 Diagram of how the MEP can be included in the achievable $Re\{Z_{MN}\}$ interval by detuning the Tx-circuit

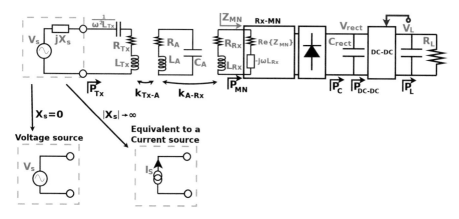

Fig. 7.20 Schematic diagram for a postregulated 3-coil link with non-resonant Tx-circuit

7.5.3   3-Coil Links

The schematic circuit of a 3-coil postregulated link is presented in Fig. 7.20. In this section, instead of analyzing the voltage and current sources separately, the equivalent Thévenin model of the driver is considered, as shown in Fig. 7.20. As discussed in the previous section while analyzing the 2-coil link, the voltage resonant Tx-circuit corresponds to the case with $X_S = 0$, and the current driver corresponds to the case where $|X_S| \rightarrow \infty$, and V_S also tends to infinity with $V_S/X_S = \text{constant} = I_S$.

The analysis of this 3-coil link is analogous to the 2-coil but using the expressions of P_C, η_{Link}, and the $Re\{Z_{MN}\}$ which maximizes them, for a 3-coil link.

In this 3-coil case, P_C is the one presented in (7.14), where the parameters $\mathbb{K}_{Tx\text{-}A}$, $\mathbb{K}_{A\text{-}Rx}$, and $\mathbb{K}_{A\text{-}Rx\text{-}L}$, were defined in (7.15) to simplify the expressions. The $Re\{Z_{MN}\}$ that maximizes P_C is presented in (7.16). The deduction of (7.14) and (7.16) can be found in Appendix C.2:

$$P_C = \frac{Q_{Rx\text{-}L}}{Q_L}\,\frac{\eta_{rect}\,\mathbb{K}_{Tx\text{-}A}\,\mathbb{K}_{A\text{-}Rx\text{-}L}}{\left(\dfrac{X_S}{R_{Tx}}\right)^2 (\mathbb{K}_{A\text{-}Rx\text{-}L}+1)^2 + (\mathbb{K}_{Tx\text{-}A}+\mathbb{K}_{A\text{-}Rx\text{-}L}+1)^2}\,\frac{V_S^2}{2R_{Tx}} \qquad (7.14)$$

$$\mathbb{K}_{Tx\text{-}A} = k_{Tx\text{-}A}{}^2 Q_{Tx} Q_A \; ; \quad \mathbb{K}_{A\text{-}Rx} = k_{A\text{-}Rx}{}^2 Q_A Q_{Rx} \; ; \quad \mathbb{K}_{A\text{-}Rx\text{-}L} = k_{A\text{-}Rx}{}^2 Q_A Q_{Rx\text{-}L} \qquad (7.15)$$

$$Re\{Z_{MN_{opt\text{-}P_{MN}}}\} = R_{Rx}\sqrt{\frac{(X_S/R_{Tx})^2(1+\mathbb{K}_{A\text{-}Rx})^2 + (\mathbb{K}_{Tx\text{-}A}+\mathbb{K}_{A\text{-}Rx}+1)^2}{(X_S/R_{Tx})^2 + (\mathbb{K}_{Tx\text{-}A}+1)^2}}. \qquad (7.16)$$

Changing X_S, the range of values of $Re\{Z_{MN_{opt\text{-}P_{MN}}}\}$ is

$$(7.12) \Rightarrow Re\{Z_{MN_{opt\text{-}PMN}}\} \in \left[\underbrace{R_{Rx}\left(\frac{\mathbb{K}_{A\text{-}Rx} + \mathbb{K}_{Tx\text{-}A} + 1}{\mathbb{K}_{Tx\text{-}A} + 1}\right)}, \underbrace{R_{Rx}(\mathbb{K}_{A\text{-}Rx} + 1)} \right)$$

$$\lim_{X_S \to 0} Re\{Z_{MN_{opt\text{-}PMN}}\} \underline{\hspace{5cm}}$$

$$\lim_{|X_S| \to \infty} Re\{Z_{MN_{opt\text{-}PMN}}\} \underline{\hspace{3cm}}$$

$$(7.17)$$

The efficiency of the 3-coil link was already calculated in Sect. 2.2.1, and the $Re\{Z_{MN}\}$ which maximizes it by achieving the MEP was deduced in Sect. 5.5. It is also presented in (7.18) using the parameters defined in (7.15):

$$Re\{Z_{MN_{opt\text{-}\eta}}\} = R_{Rx}\sqrt{\frac{(\mathbb{K}_{A\text{-}Rx} + 1)(\mathbb{K}_{A\text{-}Rx} + \mathbb{K}_{Tx\text{-}A} + 1)}{\mathbb{K}_{Tx\text{-}A} + 1}} \tag{7.18}$$

In this 3-coil case, the $Re\{Z_{MN_{opt\text{-}PMN}}\}$ obtained with a voltage source ($X_S = 0$) is less than $Re\{Z_{MN_{opt\text{-}\eta}}\}$, instead of being greater, as it was in the 2-coil case. Moreover, the $Re\{Z_{MN_{opt\text{-}PMN}}\}$ obtained with a current source ($|X_S| \to \infty$) is greater than $Re\{Z_{MN_{opt\text{-}\eta}}\}$, instead of being less. This is represented in (7.19) and proved in Appendix C.3. As a result, the range of values of $Re\{Z_{MN}\}$ and the attainability of the MEP are different from the 2-coil case, as is depicted in Fig. 7.21. Table 7.5 summarizes the MEP attainability for postregulated 3-coil links.

Voltage source

series resonant Tx

$(X_S = 0)$

Current source

series resonant Tx

$(|X_S| \to \infty)$

$$\underbrace{R_{Rx}\left(\frac{\mathbb{K}_{A\text{-}Rx} + \mathbb{K}_{Tx\text{-}A} + 1}{\mathbb{K}_{Tx\text{-}A} + 1}\right)}_{Re\{Z_{MN_{opt\text{-}PMN}}\}} < \underbrace{R_{Rx}\sqrt{\frac{(\mathbb{K}_{A\text{-}Rx} + 1)(\mathbb{K}_{A\text{-}Rx} + \mathbb{K}_{Tx\text{-}A} + 1)}{\mathbb{K}_{Tx\text{-}A} + 1}}}_{Re\{Z_{MN_{opt\text{-}\eta}}\}} < \underbrace{R_{Rx}(\mathbb{K}_{A\text{-}Rx} + 1)}_{Re\{Z_{MN_{opt\text{-}PMN}}\}}$$

$$(7.19)$$

7.5.4 N-Coil Links

The attainability of the MEP in postregulated 2-coil and 3-coil links was studied in Sects. 7.5.2 and 7.5.3, respectively, and the results for each section were summarized in Tables 7.3 and 7.5. Comparing Tables 7.3 and 7.5, it can be seen that in the cases

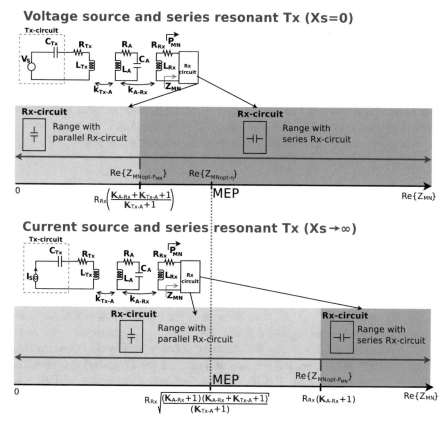

Fig. 7.21 Range of values of $Re\{Z_{MN}\}$ in a 3-coil link, for series Rx matching network, parallel Rx matching network, voltage Tx driver, and current Tx driver

Table 7.5 Summary of different cases studied for the 3-coil link, indicating whether MEP tracking is possible by adjusting the Tx-circuit amplitude

Case	Tx-circuit	Rx matching network	Attainability of MEP
I	Voltage source ($X_S = 0$)	Series	YES
II	Voltage source ($X_S = 0$)	Parallel	NO
III	Current source ($|X_S| \rightarrow \infty$)	Series	NO
IV	Current source ($|X_S| \rightarrow \infty$)	Parallel	YES

where the 2-coil link is able to achieve the MEP, the 3-coil link is not, and vice versa. This difference in the MEP attainability lies in the relative order between $Re\{Z_{MN_{opt-\eta}}\}$ and $Re\{Z_{MN_{opt-P_{MN}}}\}$ which is different in the 2-coil case (7.10) and in the 3-coil case (7.19). The influence of the relative order between $Re\{Z_{MN_{opt-\eta}}\}$

and $Re\{Z_{MN_{opt\text{-}P_{MN}}}\}$ in the MEP attainability was already graphically illustrated in Figs. 7.14 and 7.21.

It can be proved that the relative order between $Re\{Z_{MN_{opt\text{-}\eta}}\}$ and $Re\{Z_{MN_{opt\text{-}P_{MN}}}\}$ depends on the parity of the number of coils used in the link, as presented in Table 7.6. The proof of Table 7.6 can be found in Appendix C.4.

Therefore, the MEP attainability, in a link with an even number of coils, is equal to a 2-coil link which was already studied in Sect. 7.5.2, while in a link with an odd number of coils, it is equal to a 3-coil link, studied in Sect. 7.5.3.

To conclude Sect. 7.5, the MEP attainability is summarized in Table 7.7.

7.5.5   Measurement Results

In this section, some of the cases presented in Table 7.7 are implemented, using Commercially-Available Off-the-Shelf (COTS) components, and measured. The system, which operates at 13.56 MHz and is intended to power a 50 mW load, is presented in Fig. 7.22. All the parameters and component values are presented in Table 7.8.

The Tx voltage V_S was swept in order to reach the MEP. Six different cases were measured while tuning and detuning the Tx in 2-, 3-, and 4-coil links. In all cases, a parallel capacitor in the Rx matching network was used. The same type of coil (W7002, Pulse-Larsen Antennas) was used for the Tx and all the additional coils, as shown in Fig. 7.22. The Rx coil was implemented using a W7001, also from Pulse-Larsen Antennas. The total efficiency η_{TOT} as a function of V_S for the six cases measured are presented in Fig. 7.23. The power derived from source, P_S,

Table 7.6 Summary of relative order between $Re\{Z_{MN_{opt\text{-}\eta}}\}$ and $Re\{Z_{MN_{opt\text{-}P_{MN}}}\}$

N even	$\lim_{|X_S|\to\infty} Re\{Z_{MN_{opt\text{-}P_{MN}}}\} < Re\{Z_{MN_{opt\text{-}\eta}}\} < \lim_{X_S\to 0} Re\{Z_{MN_{opt\text{-}P_{MN}}}\}$
N odd	$\lim_{X_S\to 0} Re\{Z_{MN_{opt\text{-}P_{MN}}}\} < Re\{Z_{MN_{opt\text{-}\eta}}\} < \lim_{|X_S|\to\infty} Re\{Z_{MN_{opt\text{-}P_{MN}}}\}$

Table 7.7 Summary of different cases studied indicating whether MEP tracking is possible by adjusting the driver amplitude in a postregulated link

	Tx-circuit: Driver	N	Rx	MEP
	Type & resonance, X_S	(number of coils)	Matching network	Attainable?
I	Voltage ($X_S = 0$)	Even	Series	NO
II	Voltage ($X_S = 0$)	Even	Parallel	YES
III	Current ($|X_S| \to \infty$)	Even	Series	YES
IV	Current ($|X_S| \to \infty$)	Even	Parallel	NO
V	Voltage ($X_S = 0$)	Odd	Series	YES
VI	Voltage ($X_S = 0$)	Odd	Parallel	NO
VII	Current ($|X_S| \to \infty$)	Odd	Series	NO
VIII	Current ($|X_S| \to \infty$)	Odd	Parallel	YES

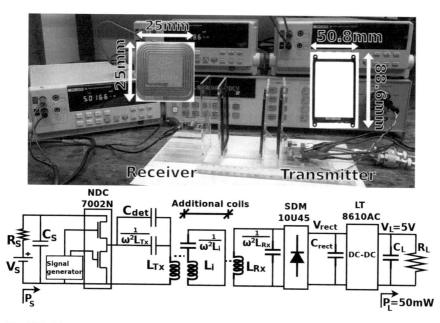

Fig. 7.22 Measurement setup and schematic diagram of the WPT link

Table 7.8 Parameters and component values for the system presented in Fig. 7.22

Parameters		Value
Tx and additional coil W7002 Pulse-Larsen Antennas Vancouver, WA 98683, United States	$L_{Tx} = L_A$	1.18 μH
	$Q_{Tx} = Q_A$	146.5
Rx W7001 Pulse-Larsen Antennas Vancouver, WA 98683, United States	L_{Rx}	877 nH
	Q_{Rx}	34.1
Carrier frequency f_S		13.56 MHz
Load power P_L		50 mW
Tx Driver NDC7002N ON semiconductor Phoenix, Arizona, United States	η_{Tx}	90%
Rectifier full wave SDM10U45 Diodes Incorporated Plano, Texas, United States	η_{rect}	90%
Regulator LT8610AC Linear Technology Norwood, Massachusetts, United States	$\eta_{DC\text{-}DC}$	75%

was determined by measuring the current through a shunt resistor (R_S, shown in Fig. 7.22) and the voltage delivered. The total system efficiency was calculated as $\eta_{TOT} = (V_L^2/R_L)/P_S$.

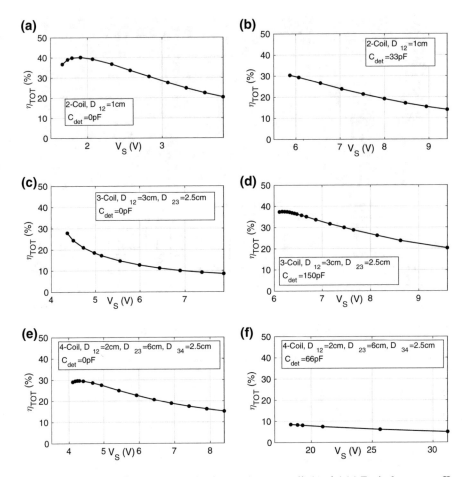

Fig. 7.23 Measured efficiency. Dij is the distance between coils i and j (**a**) Equivalent to case II Table 7.7 (**b**) Equivalent to case IV Table 7.7 (**c**) Equivalent to case VI Table 7.7 (**d**) Equivalent to case VIII Table 7.7 (**e**) Equivalent to case II Table 7.7 (**f**) Equivalent to case IV Table 7.7

Each plot in Fig. 7.23 indicates the number of coils used in the link, the distances between coils, and the value of C_{det} used to detune the Tx (Fig. 7.22). A $C_{det} = 0$ pF means that the Tx is at resonance. Additionally, the corresponding case of Table 7.7 is indicated in the caption.

The value of C_{det} was selected to generate an $|X_S|$ high enough to change the relative order between $Re\{Z_{MN_{opt-\eta}}\}$ and $Re\{Z_{MN_{opt-P_{MN}}}\}$ as previously discussed. For instance, in the case of a 2-coil link, $C_{det} = 33$ pF generates an $X_S \simeq 22\ \Omega$ at 13.56 MHz. Evaluating (5.4) and (7.12), it is deduced that $Re\{Z_{MN_{opt-P_{MN}}}\}|@X_S=22\ \Omega < Re\{Z_{MN_{opt-\eta}}\} < Re\{Z_{MN_{opt-P_{MN}}}\}|@X_S=0$. Therefore, the relative order between $Re\{Z_{MN_{opt-\eta}}\}$ and $Re\{Z_{MN_{opt-P_{MN}}}\}$ was changed by the inclusion of C_{det}. The same change could be achieved by changing the driver from voltage to current mode as was discussed.

In all the cases, V_S was decreased until the Rx DC-DC converter stopped working. The DC-DC converter could stop working for two reasons: (1) V_{rect} is below the DC-DC converter input range. (2) V_S is too low for P_C to surpass P_{DC-DC} (see Fig. 7.7), and thus a viable operating point that can deliver sufficient power to the load does not exist. The first reason can be solved by selecting another regulator. The value of V_{rect} was measured, and it was verified that V_{rect} was always in the range tolerated by the DC-DC converter, and it was not limiting the experiment. Consequently, the minimum V_S in the sweep was determined in all the cases, due to the second reason.

It can be seen in Fig. 7.23 that in the cases where it is possible to achieve the MEP, a maximum appears. In the cases of 2-coil and 4-coil, the maximum efficiency achieved is higher when the Tx is resonating as predicted by Table 7.7. However, in the case of a 3-coil link, MEP is attainable with the Tx detuned. As can be seen in Fig. 7.23c, d, by detuning the Tx in the 3-coil link, MEP tracking was possible, and the efficiency achieved increased from 28% to 37.5%.

7.5.6   Concluding Remarks

In Sect. 7.5, it was shown that the MEP attainability in postregulated links depends on (1) the type of Tx driver used (voltage or current source), (2) the resonance of the Tx-circuit, (3) the parity of the number of coils used in the inductive link, and (4) the Rx matching network (series or parallel). All these cases were summarized in Table 7.7.

It was highlighted that the Tx-circuit should be selected to operate in the cases of Table 7.7 where the MEP can be achieved. This can be done by correctly selecting the driver (current or voltage) or correctly selecting the tune (resonant or non-resonant Tx).

In the measurement results using COTS, the efficiency of a 3-coil link was improved from 28% to 37.5% by detuning the Tx-circuit, i.e., by moving from case VI to case VIII of Table 7.7, proving the importance of taking into account the analysis presented in this section.

Appendices

C.1   Deduction of (7.11) and (7.12)

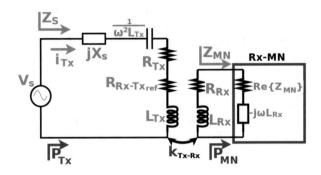

P_{MN} can be calculated as $P_{Tx} \times \eta_{Link}$

$$P_{MN} = \overbrace{\frac{1}{2} \underbrace{\left(\frac{V_S}{|Z_S|}\right)^2}_{i_{Tx}{}^2} Re\{Z_S\}}^{P_{Tx}} \times \eta_{Link}$$

$$= \overbrace{\frac{1}{2} \underbrace{\left(\frac{V_S}{|R_{Tx} + R_{Rx\text{-}Tx_{ref}} + jX_S|}\right)^2}_{i_{Tx}{}^2} (R_{Tx} + R_{Rx\text{-}Tx_{ref}})}^{P_{Tx}} \times \eta_{Link}$$

$$= \frac{V_S{}^2}{2} \frac{R_{Tx} + R_{Rx\text{-}Tx_{ref}}}{(R_{Tx} + R_{Rx\text{-}Tx_{ref}})^2 + X_S{}^2} \times \eta_{Link} \Rightarrow$$

Substituting $R_{Rx\text{-}Tx_{ref}}$ from (2.10) $\Rightarrow$

$$= \frac{V_S{}^2}{2} \frac{R_{Tx} + k_{Tx\text{-}Rx}^2 Q_{Tx} Q_{Rx\text{-}L} R_{Tx}}{(R_{Tx} + k_{Tx\text{-}Rx}^2 Q_{Tx} Q_{Rx\text{-}L} R_{Tx})^2 + X_S{}^2} \times \eta_{Link} \Rightarrow$$

$$= \frac{V_S{}^2}{2R_{Tx}} \frac{1 + k_{Tx\text{-}Rx}^2 Q_{Tx} Q_{Rx\text{-}L}}{(1 + k_{Tx\text{-}Rx}^2 Q_{Tx} Q_{Rx\text{-}L})^2 + \left(\frac{X_S}{R_{Tx}}\right)^2} \times \eta_{Link} \Rightarrow$$

Substituting η_{Link} from (2.13) $\Rightarrow$

$$= \frac{V_S^2}{2R_{Tx}} \frac{1 + \cancel{k_{Tx\text{-}Rx}^2 Q_{Tx} Q_{Rx\text{-}L}}}{(1 + k_{Tx\text{-}Rx}^2 Q_{Tx} Q_{Rx\text{-}L})^2 + \left(\frac{X_S}{R_{Tx}}\right)^2} \frac{Q_{Rx\text{-}L}}{Q_L} \frac{k_{Tx\text{-}Rx}^2 Q_{Tx} Q_{Rx\text{-}L}}{(\cancel{k_{Tx\text{-}Rx}^2 Q_{Tx} Q_{Rx\text{-}L}} + 1)}$$

$$= \frac{Q_{Rx\text{-}L}}{Q_L} \frac{k_{Tx\text{-}Rx}^2 Q_{Tx} Q_{Rx\text{-}L}}{(1 + k_{Tx\text{-}Rx}^2 Q_{Tx} Q_{Rx\text{-}L})^2 + \left(\frac{X_S}{R_{Tx}}\right)^2} \frac{V_S^2}{2R_{Tx}} \Rightarrow$$

$$P_C = \underbrace{\frac{Q_{Rx\text{-}L}}{Q_L} \frac{k_{Tx\text{-}Rx}^2 Q_{Tx} Q_{Rx\text{-}L}}{(1 + k_{Tx\text{-}Rx}^2 Q_{Tx} Q_{Rx\text{-}L})^2 + \left(\frac{X_S}{R_{Tx}}\right)^2} \frac{V_S^2}{2R_{Tx}}}_{P_{MN}} \eta_{MN} \eta_{rect}$$

which proves (7.11). Now to prove (7.12) let us substitute $Q_{Rx\text{-}L}$ from (2.11)

$$P_C = \frac{\frac{Q_{Rx} Q_L}{Q_{Rx} + Q_L}}{\cancel{Q_L}} \frac{k_{Tx\text{-}Rx}^2 Q_{Tx} \frac{Q_{Rx} Q_L}{Q_{Rx} + Q_L}}{(1 + k_{Tx\text{-}Rx}^2 Q_{Tx} \frac{Q_{Rx} Q_L}{Q_{Rx} + Q_L})^2 + \left(\frac{X_S}{R_{Tx}}\right)^2} \frac{V_S^2}{2R_{Tx}} \eta_{MN} \eta_{rect}$$

$$P_C = \frac{\overbrace{k_{Tx\text{-}Rx}^2 Q_{Tx} Q_{Rx}^2 \frac{V_S^2}{2R_{Tx}} \eta_{MN} \eta_{rect}}^{\text{independent of } Q_L} Q_L}{(Q_{Rx} + Q_L + k_{Tx\text{-}Rx}^2 Q_{Tx} Q_{Rx} Q_L)^2 + (Q_{Rx} + Q_L)^2 \left(\frac{X_S}{R_{Tx}}\right)^2}$$

$$\frac{\partial P_C}{\partial Q_L} = 0 \quad \Leftrightarrow$$

$$0 = \left(\begin{array}{l} (Q_{Rx} + Q_L + k_{Tx\text{-}Rx}^2 Q_{Tx} Q_{Rx} Q_L)^2 + (Q_{Rx} + Q_L)^2 \left(\frac{X_S}{R_{Tx}}\right)^2 \\[2mm] - 2Q_L(Q_{Rx} + Q_L + k_{Tx\text{-}Rx}^2 Q_{Tx} Q_{Rx} Q_L)(1 + k_{Tx\text{-}Rx}^2 Q_{Tx} Q_{Rx}) \\[2mm] - 2Q_L(Q_{Rx} + Q_L)\left(\frac{X_S}{R_{Tx}}\right)^2 \end{array} \right)$$

Defining $\mathbb{K}_{Tx\text{-}Rx} = k_{Tx\text{-}Rx}^2 Q_{Tx} Q_{Rx}$

$$0 = \left(\begin{array}{l} (Q_{Rx} + Q_L + \mathbb{K}_{Tx\text{-}Rx} Q_L)^2 + (Q_{Rx} + Q_L)^2 \left(\frac{X_S}{R_{Tx}}\right)^2 \\[2mm] - 2Q_L(Q_{Rx} + Q_L + \mathbb{K}_{Tx\text{-}Rx} Q_L)(1 + \mathbb{K}_{Tx\text{-}Rx}) \\[2mm] - 2Q_L(Q_{Rx} + Q_L)\left(\frac{X_S}{R_{Tx}}\right)^2 \end{array} \right)$$

$$0 = \begin{pmatrix} [Q_{Rx} + Q_L(1 + \mathbb{K}_{Tx\text{-}Rx})]^2 \\ -2Q_L[Q_{Rx} + Q_L(1 + \mathbb{K}_{Tx\text{-}Rx})](1 + \mathbb{K}_{Tx\text{-}Rx}) \\ \left(\frac{X_S}{R_{Tx}}\right)^2 [(Q_{Rx} + Q_L)^2 - 2Q_L(Q_{Rx} + Q_L)] \end{pmatrix}$$

$$0 = \begin{pmatrix} Q_{Rx}^2 + Q_L^2(1 + \mathbb{K}_{Tx\text{-}Rx})^2 + \cancel{2Q_{Rx}Q_L(1 + \mathbb{K}_{Tx\text{-}Rx})} \\ \cancel{-2Q_{Rx}Q_L(1 + \mathbb{K}_{Tx\text{-}Rx})} - 2Q_L^2(1 + \mathbb{K}_{Tx\text{-}Rx})^2 \\ \left(\frac{X_S}{R_{Tx}}\right)^2 (Q_{Rx}^2 + Q_L^2 + \cancel{2Q_{Rx}Q_L} \cancel{-2Q_{Rx}Q_L} - 2Q_L^2) \end{pmatrix}$$

$$0 = Q_{Rx}^2 - Q_L^2(1 + \mathbb{K}_{Tx\text{-}Rx})^2 + \left(\frac{X_S}{R_{Tx}}\right)^2 (Q_{Rx}^2 - Q_L^2)$$

$$0 = Q_{Rx}^2 \left[1 + \left(\frac{X_S}{R_{Tx}}\right)^2\right] - Q_L^2 \left[(1 + \mathbb{K}_{Tx\text{-}Rx})^2 + \left(\frac{X_S}{R_{Tx}}\right)^2\right] \Rightarrow$$

$$Q_L = Q_{Rx} \sqrt{\frac{1 + \left(\frac{X_S}{R_{Tx}}\right)^2}{(1 + \mathbb{K}_{Tx\text{-}Rx})^2 + \left(\frac{X_S}{R_{Tx}}\right)^2}}$$

as $Q_L = \dfrac{\omega L_{Rx}}{Re\{Z_{MN}\}}$ and $Q_{Rx} = \dfrac{\omega L_{Rx}}{R_{Rx}} \Rightarrow$

$$\frac{\cancel{\omega L_{Rx}}}{Re\{Z_{MN}\}} = \frac{\cancel{\omega L_{Rx}}}{R_{Rx}} \sqrt{\frac{1 + \left(\frac{X_S}{R_{Tx}}\right)^2}{(1 + \mathbb{K}_{Tx\text{-}Rx})^2 + \left(\frac{X_S}{R_{Tx}}\right)^2}}$$

$$Re\{Z_{MN}\} = R_{Rx} \sqrt{\frac{(1 + \mathbb{K}_{Tx\text{-}Rx})^2 + \left(\frac{X_S}{R_{Tx}}\right)^2}{1 + \left(\frac{X_S}{R_{Tx}}\right)^2}}$$

$$Re\{Z_{MN_{opt\text{-}P_{MN}}}\} = R_{Rx} \sqrt{\frac{(1 + k_{Tx\text{-}Rx}^2 Q_{Tx} Q_{Rx})^2 + (X_S/R_{Tx})^2}{(X_S/R_{Tx})^2 + 1}}$$

C.2 Deduction of (7.14) and (7.16)

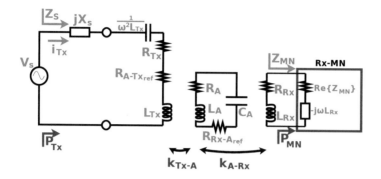

P_{MN} can be calculated as $P_{Tx} \times \eta_{Link}$

$$P_{MN} = \frac{1}{2} \underbrace{\left(\frac{V_S}{|Z_S|} \right)^2 Re\{Z_S\}}_{i_{Tx}^2} \times \eta_{Link}$$
$$\overbrace{\phantom{\frac{1}{2} \left(\frac{V_S}{|Z_S|} \right)^2 Re\{Z_S\}}}^{P_{Tx}}$$

$$= \frac{1}{2} \underbrace{\left(\frac{V_S}{|R_{Tx} + R_{A\text{-}Tx_{ref}} + jX_S|} \right)^2}_{i_{Tx}^2} (R_{Tx} + R_{A\text{-}Tx_{ref}}) \times \eta_{Link}$$

$$= \frac{V_S^2}{2} \frac{R_{Tx} + R_{A\text{-}Tx_{ref}}}{(R_{Tx} + R_{A\text{-}Tx_{ref}})^2 + X_S^2} \times \eta_{Link} \Rightarrow$$

Substituting $R_{A\text{-}Tx_{ref}}$ from (2.30) $\Rightarrow$

$$= \frac{V_S^2}{2} \frac{R_{Tx} + \frac{k_{Tx\text{-}A}^2 Q_{Tx} Q_A R_{Tx}}{k_{A\text{-}Rx}^2 Q_A Q_{Rx\text{-}L} + 1}}{\left(R_{Tx} + \frac{k_{Tx\text{-}A}^2 Q_{Tx} Q_A R_{Tx}}{k_{A\text{-}Rx}^2 Q_A Q_{Rx\text{-}L} + 1} \right)^2 + X_S^2} \times \eta_{Link} \Rightarrow$$

$$= \frac{V_S^2}{2 R_{Tx}} \frac{1 + \frac{k_{Tx\text{-}A}^2 Q_{Tx} Q_A}{k_{A\text{-}Rx}^2 Q_A Q_{Rx\text{-}L} + 1}}{\left(1 + \frac{k_{Tx\text{-}A}^2 Q_{Tx} Q_A}{k_{A\text{-}Rx}^2 Q_A Q_{Rx\text{-}L} + 1} \right)^2 + \left(\frac{X_S}{R_{Tx}} \right)^2} \times \eta_{Link} \Rightarrow$$

Defining $\mathbb{K}_{Tx\text{-}A} = k_{Tx\text{-}A}{}^2 Q_{Tx} Q_A$; $\mathbb{K}_{A\text{-}Rx\text{-}L} = k_{A\text{-}Rx}{}^2 Q_A Q_{Rx\text{-}L}$

$$= \frac{V_S^2}{2R_{Tx}} \frac{1 + \frac{\mathbb{K}_{Tx\text{-}A}}{\mathbb{K}_{A\text{-}Rx\text{-}L}+1}}{\left(1 + \frac{\mathbb{K}_{Tx\text{-}A}}{\mathbb{K}_{A\text{-}Rx\text{-}L}+1}\right)^2 + \left(\frac{X_S}{R_{Tx}}\right)^2} \times \eta_{Link} \Rightarrow$$

$$= \frac{V_S^2}{2R_{Tx}} \frac{(\mathbb{K}_{A\text{-}Rx\text{-}L} + \mathbb{K}_{Tx\text{-}A} + 1)(\mathbb{K}_{A\text{-}Rx\text{-}L} + 1)}{(\mathbb{K}_{A\text{-}Rx\text{-}L} + \mathbb{K}_{Tx\text{-}A} + 1)^2 + \left(\frac{X_S}{R_{Tx}}\right)^2 (\mathbb{K}_{A\text{-}Rx\text{-}L} + 1)^2} \times \eta_{Link}$$

Substituting η_{Link} from (2.32),

and using the defined $\mathbb{K}_{Tx\text{-}A}$ and $\mathbb{K}_{A\text{-}Rx\text{-}L}$ $\Rightarrow$

$$P_{MN} = \frac{V_S^2}{2R_{Tx}} \left(\frac{\cancel{(\mathbb{K}_{A\text{-}Rx\text{-}L} + \mathbb{K}_{Tx\text{-}A} + 1)}\,\cancel{(\mathbb{K}_{A\text{-}Rx\text{-}L} + 1)}}{(\mathbb{K}_{A\text{-}Rx\text{-}L} + \mathbb{K}_{Tx\text{-}A} + 1)^2 + \left(\frac{X_S}{R_{Tx}}\right)^2 (\mathbb{K}_{A\text{-}Rx\text{-}L} + 1)^2} \times \dots \right.$$
$$\left. \frac{Q_{Rx\text{-}L}}{Q_L} \frac{\mathbb{K}_{A\text{-}Rx\text{-}L}}{\cancel{(\mathbb{K}_{A\text{-}Rx\text{-}L} + 1)}} \frac{\mathbb{K}_{Tx\text{-}A}}{\cancel{(\mathbb{K}_{Tx\text{-}A} + \mathbb{K}_{A\text{-}Rx\text{-}L} + 1)}} \right)$$

$$= \frac{Q_{Rx\text{-}L}}{Q_L} \left(\frac{\mathbb{K}_{Tx\text{-}A}\mathbb{K}_{A\text{-}Rx\text{-}L}}{(\mathbb{K}_{A\text{-}Rx\text{-}L} + \mathbb{K}_{Tx\text{-}A} + 1)^2 + \left(\frac{X_S}{R_{Tx}}\right)^2 (\mathbb{K}_{A\text{-}Rx\text{-}L} + 1)^2} \right) \frac{V_S^2}{2R_{Tx}} \Rightarrow$$

$$P_C = \underbrace{\frac{Q_{Rx\text{-}L}}{Q_L} \left(\frac{\mathbb{K}_{Tx\text{-}A}\mathbb{K}_{A\text{-}Rx\text{-}L}}{(\mathbb{K}_{A\text{-}Rx\text{-}L} + \mathbb{K}_{Tx\text{-}A} + 1)^2 + \left(\frac{X_S}{R_{Tx}}\right)^2 (\mathbb{K}_{A\text{-}Rx\text{-}L} + 1)^2} \right) \frac{V_S^2}{2R_{Tx}}}_{P_{MN}} \eta_{rect}$$

which proves (7.14). Now to prove (7.16), let us substitute $Q_{Rx\text{-}L}$ from (2.11); to do this note that

$$\mathbb{K}_{A\text{-}Rx\text{-}L} = k_{A\text{-}Rx}{}^2 Q_A Q_{Rx\text{-}L} = \underbrace{k_{A\text{-}Rx}{}^2 Q_A Q_{Rx}}_{\mathbb{K}_{A\text{-}Rx}} \frac{Q_L}{Q_L + Q_{Rx}} \Rightarrow$$

$$P_C = \left(\frac{\mathbb{K}_{Tx\text{-}A}\mathbb{K}_{A\text{-}Rx}\frac{Q_L}{Q_L+Q_{Rx}}}{\left(\mathbb{K}_{A\text{-}Rx}\frac{Q_L}{Q_L+Q_{Rx}} + \mathbb{K}_{Tx\text{-}A} + 1\right)^2 + \left(\frac{X_S}{R_{Tx}}\right)^2 \left(\mathbb{K}_{A\text{-}Rx}\frac{Q_L}{Q_L+Q_{Rx}} + 1\right)^2} \right) \times \dots$$

$$\times \frac{Q_{Rx}}{Q_L + Q_{Rx}} \frac{V_S^2}{2R_{Tx}} \eta_{rect}$$

$$P_C = \left(\cfrac{Q_L}{\left(\begin{array}{c} \left[\mathbb{K}_{A\text{-}Rx}\, Q_L + (\mathbb{K}_{Tx\text{-}A}+1)(Q_L+Q_{Rx}) \right]^2 + \ldots \\[2mm] + \left(\cfrac{X_S}{R_{Tx}} \right)^2 (\mathbb{K}_{A\text{-}Rx}\, Q_L + Q_L + Q_{Rx})^2 \end{array} \right)} \right) \times \ldots$$

$$\ldots \times \mathbb{K}_{Tx\text{-}A}\, \mathbb{K}_{A\text{-}Rx}\, Q_{Rx} \frac{V_S^2}{2R_{Tx}} \eta_{rect}$$

$$P_C = \left(\cfrac{Q_L}{\left(\begin{array}{c} \left[Q_L(\mathbb{K}_{A\text{-}Rx} + \mathbb{K}_{Tx\text{-}A} + 1) + Q_{Rx}(\mathbb{K}_{Tx\text{-}A} + 1) \right]^2 + \ldots \\[2mm] + \left(\cfrac{X_S}{R_{Tx}} \right)^2 [Q_L(\mathbb{K}_{A\text{-}Rx} + 1) + Q_{Rx}]^2 \end{array} \right)} \right) \times \ldots$$

$$\ldots \times \underbrace{\mathbb{K}_{Tx\text{-}A}\, \mathbb{K}_{A\text{-}Rx}\, Q_{Rx} \frac{V_S^2}{2R_{Tx}} \eta_{rect}}_{\text{independent of } Q_L}$$

$$\frac{\partial P_C}{\partial Q_L} = 0 \quad \Leftrightarrow$$

$$0 = \left(\begin{array}{c} \left[Q_L(\mathbb{K}_{A\text{-}Rx} + \mathbb{K}_{Tx\text{-}A} + 1) + Q_{Rx}(\mathbb{K}_{Tx\text{-}A} + 1) \right]^2 + \ldots \\[2mm] \ldots + \left(\cfrac{X_S}{R_{Tx}} \right)^2 [Q_L(\mathbb{K}_{A\text{-}Rx} + 1) + Q_{Rx}]^2 \ldots \\[2mm] -2Q_L \left[Q_L(\mathbb{K}_{A\text{-}Rx} + \mathbb{K}_{Tx\text{-}A} + 1) + Q_{Rx}(\mathbb{K}_{Tx\text{-}A} + 1) \right] (\mathbb{K}_{A\text{-}Rx} + \mathbb{K}_{Tx\text{-}A} + 1) \\[2mm] -2Q_L \left(\cfrac{X_S}{R_{Tx}} \right)^2 [Q_L(\mathbb{K}_{A\text{-}Rx} + 1) + Q_{Rx}](\mathbb{K}_{A\text{-}Rx} + 1) \end{array} \right)$$

$$0 = \left(\begin{array}{c} Q_L^2(\mathbb{K}_{A\text{-}Rx} + \mathbb{K}_{Tx\text{-}A} + 1)^2 + Q_{Rx}^2(\mathbb{K}_{Tx\text{-}A} + 1)^2 \dots \\ \overline{+2Q_L(\mathbb{K}_{A\text{-}Rx} + \mathbb{K}_{Tx\text{-}A} + 1)Q_{Rx}(\mathbb{K}_{Tx\text{-}A} + 1)} \\ -2Q_L^2(\mathbb{K}_{A\text{-}Rx} + \mathbb{K}_{Tx\text{-}A} + 1)^2 \dots \\ \overline{-2Q_L Q_{Rx}(\mathbb{K}_{Tx\text{-}A} + 1)(\mathbb{K}_{A\text{-}Rx} + \mathbb{K}_{Tx\text{-}A} + 1)} \\ +\left(\dfrac{X_S}{R_{Tx}}\right)^2 \left[Q_L^2(\mathbb{K}_{A\text{-}Rx} + 1)^2 + Q_{Rx}^2 + 2Q_L(\mathbb{K}_{A\text{-}Rx} + 1)Q_{Rx} \right] \dots \\ -\left(\dfrac{X_S}{R_{Tx}}\right)^2 [2Q_L^2(\mathbb{K}_{A\text{-}Rx} + 1)^2 + 2Q_L Q_{Rx}(\mathbb{K}_{A\text{-}Rx} + 1)] \dots \end{array} \right)$$

$$0 = \left(\begin{array}{c} -Q_L^2(\mathbb{K}_{A\text{-}Rx} + \mathbb{K}_{Tx\text{-}A} + 1)^2 + Q_{Rx}^2(\mathbb{K}_{Tx\text{-}A} + 1)^2 \dots \\ +\left(\dfrac{X_S}{R_{Tx}}\right)^2 \left[-Q_L^2(\mathbb{K}_{A\text{-}Rx} + 1)^2 + Q_{Rx}^2 \right] \dots \end{array} \right)$$

$$0 = \left(\begin{array}{c} -Q_L^2 \left[(\mathbb{K}_{A\text{-}Rx} + \mathbb{K}_{Tx\text{-}A} + 1)^2 + \left(\dfrac{X_S}{R_{Tx}}\right)^2 (\mathbb{K}_{A\text{-}Rx} + 1)^2 \right] \\ +Q_{Rx}^2 \left[(\mathbb{K}_{Tx\text{-}A} + 1)^2 + \left(\dfrac{X_S}{R_{Tx}}\right)^2 \right] \end{array} \right) \Rightarrow$$

$$Q_L = Q_{Rx} \sqrt{ \dfrac{(\mathbb{K}_{Tx\text{-}A} + 1)^2 + \left(\dfrac{X_S}{R_{Tx}}\right)^2}{(\mathbb{K}_{A\text{-}Rx} + \mathbb{K}_{Tx\text{-}A} + 1)^2 + \left(\dfrac{X_S}{R_{Tx}}\right)^2 (\mathbb{K}_{A\text{-}Rx} + 1)^2} }$$

as $Q_L = \dfrac{\omega L_{Rx}}{Re\{Z_{MN}\}}$ and $Q_{Rx} = \dfrac{\omega L_{Rx}}{R_{Rx}} \Rightarrow$

$$Re\{Z_{MN_{opt\text{-}P_{MN}}}\} = R_{Rx} \sqrt{ \dfrac{(\mathbb{K}_{A\text{-}Rx} + \mathbb{K}_{Tx\text{-}A} + 1)^2 + \left(\dfrac{X_S}{R_{Tx}}\right)^2 (\mathbb{K}_{A\text{-}Rx} + 1)^2}{(\mathbb{K}_{Tx\text{-}A} + 1)^2 + \left(\dfrac{X_S}{R_{Tx}}\right)^2} }$$

C.3   Proof of (7.19)

$$\cancel{R_{Rx}}\left(\frac{\mathbb{K}_{A\text{-}Rx} + \mathbb{K}_{Tx\text{-}A} + 1}{\mathbb{K}_{Tx\text{-}A} + 1}\right) \overset{?}{<} \cancel{R_{Rx}}\sqrt{\frac{(\mathbb{K}_{A\text{-}Rx} + 1)(\mathbb{K}_{A\text{-}Rx} + \mathbb{K}_{Tx\text{-}A} + 1)}{\mathbb{K}_{Tx\text{-}A} + 1}} \Leftrightarrow$$

$$\left(\frac{\mathbb{K}_{A\text{-}Rx} + \mathbb{K}_{Tx\text{-}A} + 1}{\mathbb{K}_{Tx\text{-}A} + 1}\right)^{\cancel{2}} \overset{?}{<} \frac{(\mathbb{K}_{A\text{-}Rx} + 1)(\cancel{\mathbb{K}_{A\text{-}Rx} + \mathbb{K}_{Tx\text{-}A} + 1})}{\cancel{\mathbb{K}_{Tx\text{-}A} + 1}} \Leftrightarrow$$

$$\mathbb{K}_{A\text{-}Rx} + \mathbb{K}_{Tx\text{-}A} + 1 \overset{?}{<} (\mathbb{K}_{A\text{-}Rx} + 1)(\mathbb{K}_{Tx\text{-}A} + 1) \Leftrightarrow$$

$$\cancel{\mathbb{K}_{A\text{-}Rx}} + \cancel{\mathbb{K}_{Tx\text{-}A}} + \cancel{1} \overset{?}{<} \mathbb{K}_{A\text{-}Rx}\mathbb{K}_{Tx\text{-}A} + \cancel{\mathbb{K}_{A\text{-}Rx}} + \cancel{\mathbb{K}_{Tx\text{-}A}} + \cancel{1} \Leftrightarrow$$

$$0 < \mathbb{K}_{A\text{-}Rx}\mathbb{K}_{Tx\text{-}A}$$

both $\mathbb{K}_{A\text{-}Rx}$ and $\mathbb{K}_{Tx\text{-}A}$ are greater than zero,

which proves the first part of the inequation. To prove the second part

$$\cancel{R_{Rx}}\sqrt{\frac{(\mathbb{K}_{A\text{-}Rx} + 1)(\mathbb{K}_{A\text{-}Rx} + \mathbb{K}_{Tx\text{-}A} + 1)}{\mathbb{K}_{Tx\text{-}A} + 1}} \overset{?}{<} \cancel{R_{Rx}}(\mathbb{K}_{A\text{-}Rx} + 1) \Leftrightarrow$$

$$\frac{(\cancel{\mathbb{K}_{A\text{-}Rx} + 1})(\mathbb{K}_{A\text{-}Rx} + \mathbb{K}_{Tx\text{-}A} + 1)}{\mathbb{K}_{Tx\text{-}A} + 1} \overset{?}{<} (\mathbb{K}_{A\text{-}Rx} + 1)^{\cancel{2}} \Leftrightarrow$$

$$\mathbb{K}_{A\text{-}Rx} + \mathbb{K}_{Tx\text{-}A} + 1 \overset{?}{<} (\mathbb{K}_{A\text{-}Rx} + 1)(\mathbb{K}_{Tx\text{-}A} + 1) \Leftrightarrow$$

$$\cancel{\mathbb{K}_{A\text{-}Rx}} + \cancel{\mathbb{K}_{Tx\text{-}A}} + \cancel{1} \overset{?}{<} \mathbb{K}_{A\text{-}Rx}\mathbb{K}_{Tx\text{-}A} + \cancel{\mathbb{K}_{A\text{-}Rx}} + \cancel{\mathbb{K}_{Tx\text{-}A}} + \cancel{1} \Leftrightarrow$$

$$0 < \mathbb{K}_{A\text{-}Rx}\mathbb{K}_{Tx\text{-}A}$$

both $\mathbb{K}_{A\text{-}Rx}$ and $\mathbb{K}_{Tx\text{-}A}$ are greater than zero,

which proves the second part of the inequation.

C.4   Deduction of Table 7.6

The rectifier output power, P_C, can be calculated as the product of $P_{Tx}, \eta_{Link}, \eta_{MN}$, and η_{rect}:

$$P_C = \underbrace{\eta_{MN} \eta_{rect}}_{} \; \eta_{Link} \; P_{Tx} \; .$$

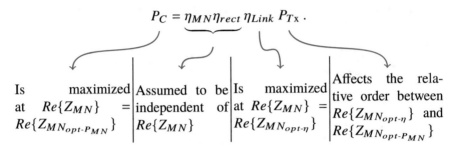

| Is maximized at $Re\{Z_{MN}\}$ $=$ $Re\{Z_{MN_{opt\text{-}PMN}}\}$ | Assumed to be independent of $Re\{Z_{MN}\}$ | Is maximized at $Re\{Z_{MN}\}$ $=$ $Re\{Z_{MN_{opt\text{-}\eta}}\}$ | Affects the relative order between $Re\{Z_{MN_{opt\text{-}\eta}}\}$ and $Re\{Z_{MN_{opt\text{-}PMN}}\}$ |

Therefore, the relative order between $Re\{Z_{MN_{opt\text{-}\eta}}\}$ and $Re\{Z_{MN_{opt\text{-}PMN}}\}$, which is what we are analyzing in this section, depends on P_{Tx}. To calculate P_{Tx} in an N-coil link, consider the system shown in Fig. C.1. The R_{Li} is the reflected resistance in coil i from coil $i+1$, which was deduced in (2.37):

$$R_{Li} = \frac{\omega^2 k_{i\text{-}(i+1)}{}^2 L_i L_{(i+1)}}{R_{(i+1)} + R_{L(i+1)}}. \tag{7.20}$$

The P_{Tx} can be calculated as

$$P_{Tx} = \frac{1}{2} \underbrace{\left(\frac{V_S}{|Z_S|}\right)^2}_{i_{Tx}{}^2} Re\{Z_S\} = \frac{V_S^2}{2} \frac{Re\{Z_S\}}{|Z_S|^2}$$

$$= \frac{V_S^2}{2} \frac{(R_{Tx} + R_{L1})}{X_S^2 + (R_{Tx} + R_{L1})^2} \tag{7.21}$$

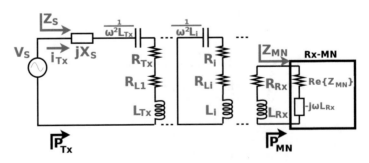

Fig. C.1 Lumped circuit model of an N-coil link. The R_{Li} is the reflected resistance in coil i from coil $i+1$

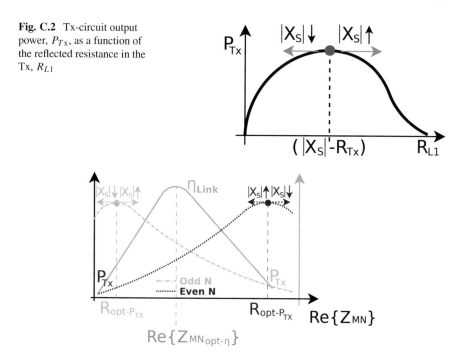

Fig. C.2 Tx-circuit output power, P_{Tx}, as a function of the reflected resistance in the Tx, R_{L1}

Fig. C.3 P_{Tx} and η_{Link} as a funtion of $Re\{Z_{MN}\}$

where V_S is the peak voltage of the sinusoidal source. From (7.21), the value of R_{L1} that maximizes P_{Tx} is $R_{L1} = |X_S| - R_{Tx}$ (assuming $|X_S| > R_{Tx}$). Thus, P_{Tx} can be graphically represented as shown in Fig. C.2.

The reflected resistance, R_{Li}, is inversely proportional to the total resultant resistance in the consecutive coil (7.20). Thus, each extra coil inverts the relationship between $Re\{Z_{MN}\}$ and R_{L1}. In a link with an even number of coils, R_{L1} as a function of $Re\{Z_{MN}\}$ is monotonically decreasing, while in a link with an odd number of coils, it is monotonically increasing. Using this result, Fig. C.2 can be redone by changing the horizontal axis from R_{L1} to $Re\{Z_{MN}\}$ as it is presented in Fig. C.3. The $Re\{Z_{MN}\}$ that maximizes P_{Tx}, $R_{opt\text{-}P_{Tx}}$, and its dependence with X_S is represented in Fig. C.3. For instance, an increase in $|X_S|$ ($|X_S| \uparrow$) moves the maximum of P_{Tx} to the right in the case of an odd number of coils and to the left in the case of an even number of coils. When $|X_S| < R_{Tx}$, P_{Tx} has no relative maximum value ($\partial P_{Tx}/\partial Re\{Z_{MN}\} \neq 0 \; \forall \; Re\{Z_{MN}\} > 0$ and $\forall \; N$), and it is a monotonically increasing function if N is even and monotonically decreasing function if N is odd, being N the number of coils.

Since P_{Tx} is always positive with no relative minimum values for any X_S, it can be proved that the $Re\{Z_{MN}\}$ that maximizes the product ($P_C = P_{Tx}\eta_{Link}\eta_{MN}\eta_{rect}$), $Re\{Z_{MN_{opt\text{-}P_{MN}}}\}$, fulfills the following properties: $Re\{Z_{MN_{opt\text{-}P_{MN}}}\} < Re\{Z_{MN_{opt\text{-}\eta}}\}$ if $R_{opt\text{-}P_{Tx}} < Re\{Z_{MN_{opt\text{-}\eta}}\}$ and

$Re\{Z_{MN_{opt-P_{MN}}}\} > Re\{Z_{MN_{opt-\eta}}\}$ if $R_{opt-P_{Tx}} > Re\{Z_{MN_{opt-\eta}}\}$. Using this property, and the previous discussion of how X_S affects $R_{opt-P_{Tx}}$, Table 7.6 can be deduced.

References

1. H. Lee, An auto-reconfigurable $2 \times /4 \times$ AC-DC regulator for wirelessly powered biomedical implants with 28% link efficiency enhancement. IEEE Trans. VLSI Syst. **24**(4), 1598–1602 (2016). ISSN 1063-8210. https://doi.org/10.1109/TVLSI.2015.2452918
2. G. Wang, W. Liu, M. Sivaprakasam, G.A. Kendir, Design and analysis of an adaptive transcutaneous power telemetry for biomedical implants. IEEE Trans. Circuits Syst. I **52**(10), 2109–2117 (2005). ISSN 1549-8328. https://doi.org/10.1109/TCSI.2005.852923
3. X. Li, C. Y. Tsui, W.H. Ki, Power management analysis of inductively-powered implants with 1X/2X reconfigurable rectifier. IEEE Trans. Circuits Syst. I **62**(3), 617–624 (2015). ISSN 1549-8328. https://doi.org/10.1109/TCSI.2014.2366814
4. P. Si, A.P. Hu, J.W. Hsu, M. Chiang, Y. Wang, S. Malpas, D. Budgett, Wireless power supply for implantable biomedical device based on primary input voltage regulation, in *IEEE Conference on Industrial Electronics and Applications*, pp. 235–239 (2007). https://doi.org/10.1109/ICIEA.2007.4318406
5. J. Tian, A.P. Hu, Stabilising the output voltage of wireless power pickup through parallel tuned DC-voltage controlled variable capacitor. Electron. Lett. **52**(9):758–759 (2016)
6. B. Lee, M. Kiani, M. Ghovanloo, A triple-loop inductive power transmission system for biomedical applications. IEEE Trans. Biomed. Circuits Syst. **10**(1), 138–148 (2015)
7. P. Si, A.P. Hu, S. Malpas, D. Budgett, A frequency control method for regulating wireless power to implantable devices. IEEE Trans. Biomed. Circuits Syst. **2**(1), 22–29 (2008)
8. T.D. Yeo, D. Kwon, S.T. Khang, J.W. Yu, Design of maximum efficiency tracking control scheme for closed-loop wireless power charging system employing series resonant tank. IEEE Trans. Power Electron. **32**(1), 471–478 (2017)
9. K. Colak, E. Asa, M. Bojarski, D. Czarkowski, O.C. Onar, A novel phase-shift control of semibridgeless active rectifier for wireless power transfer. IEEE Trans. Power Electron. **30**(11), 6288–6297 (2015)
10. S. Chen, H. Li, Y. Tang, A burst mode pulse density modulation scheme for inductive power transfer systems without communication modules, in IEEE Applied Power Electronics Conference and Exposition (IEEE, 2018), pp. 1071–1075
11. R.F. Xue, K.W. Cheng, M. Je, High-efficiency wireless power transfer for biomedical implants by optimal resonant load transformation. IEEE Trans. Circuits Syst. I **60**(4), 867–874 (2013). ISSN 1549-8328. https://doi.org/10.1109/TCSI.2012.2209297
12. M. Fu, H. Yin, X. Zhu, C. Ma, Analysis and tracking of optimal load in wireless power transfer systems. IEEE Trans. Power Electron. **30**(7), 3952–3963 (2015)
13. M. Kiani, B. Lee, P. Yeon, M. Ghovanloo, A Q-modulation technique for efficient inductive power transmission. IEEE J. Solid-State Circuits **50**(12), 2839–2848 (2015). ISSN 0018-9200. https://doi.org/10.1109/JSSC.2015.2453201
14. B. Lee, P. Yeon, M. Ghovanloo, A multicycle Q-modulation for dynamic optimization of inductive links. IEEE Trans. Ind. Electron. **63**(8), 5091–5100 (2016)
15. H.S. Gougheri, M. Kiani, Current-based resonant power delivery with multi-cycle switching for extended-range inductive power transmission. IEEE Trans. Circuits Syst. I **63**(9), 1543–1552 (2016)
16. M. Fu, H. Yin, M. Liu, C. Ma, Loading and power control for a high-efficiency class E PA-driven megahertz WPT system. IEEE Trans. Ind. Electron. **63**(11), 6867–6876 (2016). ISSN 0278-0046. https://doi.org/10.1109/TIE.2016.2582733

17. H. Li, J. Li, K. Wang, W. Chen, X. Yang, A maximum efficiency point tracking control scheme for wireless power transfer systems using magnetic resonant coupling. IEEE Trans. Power Electron. **30**(7), 3998–4008 (2015). ISSN 0885-8993. https://doi.org/10.1109/TPEL.2014.2349534
18. W.X. Zhong, S.Y.R. Hui, Maximum energy efficiency tracking for wireless power transfer systems. IEEE Trans. Power Electron. **30**(7), 4025–4034 (2015)
19. Y. Narusue, Y. Kawahara, T. Asami, Maximizing the efficiency of wireless power transfer with a receiver-side switching voltage regulator. Wireless Power Transf. 1–13 (2017) ISSN 2052-8418. https://doi.org/10.1017/wpt.2016.14
20. M. Kiani, M. Ghovanloo, The circuit theory behind coupled-mode magnetic resonance-based wireless power transmission. IEEE Trans. Circuits Syst. I **59**(9), 2065–2074 (2012). ISSN 1549-8328
21. S. Stoecklin, T. Volk, A. Yousaf, L. Reindl, A maximum efficiency point tracking system for wireless powering of biomedical implants. Procedia Eng. **120**, 451–454 (2015). ISSN 1877-7058. https://doi.org/10.1016/j.proeng.2015.08.666
22. C. Florian, F. Mastri, R.P. Paganelli, D. Masotti, A. Costanzo, Theoretical and numerical design of a wireless power transmission link with GaN-based transmitter and adaptive receiver. IEEE Trans. Microw. Theory Techn. **62**(4), 931–946 (2014). ISSN 0018-9480. https://doi.org/10.1109/TMTT.2014.2303949

Chapter 8
System Design Examples

8.1 Radio Frequency Identification (RFID)

Automatic identification procedures are widely used in a variety of applications to track or provide information about people, animals, and goods [1]. Radio Frequency Identification (RFID) is a contactless identification system, which unlike the barcode labels can be programmed, has larger storage capacity, and does not require direct visual contact for reading. For that reason, RFID is currently used in many applications such as access control, transportation payments (bus, trains, or to collect tolls on highways), and animal identification. The animal identification is one of the oldest uses of RFID, where barcodes do not work well, and it is used both for pets and livestock identification.

A brief introduction to RFID was presented in Sect. 4.3.2, where the discussion was focused on back telemetry. In this chapter, we further address the RFID example presented in that section, applying some of the concepts developed throughout the book. First, in Sect. 8.1.1, the actual RFID link, which was introduced in Sect. 4.3.2, is described. Then in Sect. 8.1.2, we calculate and simulate the Power Transfer Efficiency (PTE) and Power Delivered to the Load (PDL) in charging and reading phases of RFID operation. Finally, in Sect. 8.1.3, the use of a third coil between the reader and tag to increase read distance is addressed. The performance of this 3-coil link is also both calculated and simulated, taking into account the effects of the additional coil in the tag response by Frequency Shift Keying (FSK).

8.1.1 RFID Link Introduction

The RFID system, which was introduced in Sect. 4.3.2 and described in more details in this section, is widely used for cattle identification. This particular example follows the guidelines of the International Committee for Animal Recording (ICAR)

[2]. It corresponds to the technology used in Uruguay, where the use of RFID tags in cattle is mandatory.

The reader (Tx) is a low-cost reader based on a TMS3705 transponder base station ASIC from Texas Instruments (TI). The tag (Rx) is the RI-INL-R9QM, also from TI. The reader and tag are shown in Fig. 4.9a, b, respectively.

The operation is divided into two phases: the charging phase and the reading phase. During the charging phase (50 ms), the tag receives and stores energy via an inductive WPT link operating at 134.2 kHz. During that period, no data is transferred. When the tag detects the end of the charging phase, with the absence of the power carrier, it starts the reading phase by transmitting its ID via FSK. The tag has no energy source; thus during the reading phase, it is powered from the stored energy during the charging phase. The bit "0" is represented by the same frequency as the power carrier in the charging phase (134.2 kHz), while the bit "1" is represented by a lower frequency of 123.2 kHz. The response lasts 16 cycles per bit, which takes $\simeq 120\,\mu$s/bit and $\simeq 15$ ms in total, to deliver the 112 bit data telegram as defined in ISO 11784 and ISO 11785 [2, 3].

The lumped circuit model for the inductive link in each phase is presented in Fig. 8.1 and all the component values summarized in Table 8.1.

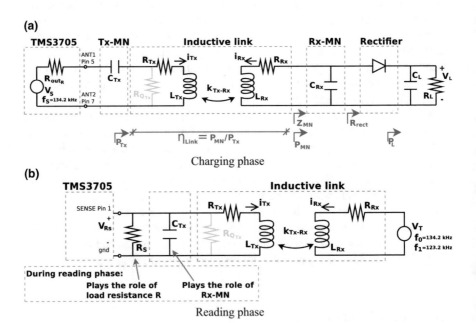

Fig. 8.1 Lumped circuit models for the charging (**a**) and reading (**b**) phases. The component values are presented in Table 8.1. $R_{Q_{Tx}}$ is added to decrease the reader coil quality factor. It is in gray because, for the sake of simplicity, it is not going to be considered at the beginning of the analysis, but it is fully considered later

Table 8.1 Component values of Fig. 8.1

Reader (Tx)	Tag (Rx)
Carrier frequency $f_s = 134.2\,\text{kHz}$	Low bit frequency $f_0 = 134.2\,\text{kHz}$
	High bit frequency $f_1 = 123.2\,\text{kHz}$
Charging phase duration 50 ms	Reading phase duration
	16 cycles ($\sim$120 μs) per bit
	112 bits, $\sim$15 ms in total
$R_{Tx} = 15.9\,\Omega$ @ 134.2 kHz	$R_{Rx} = 44.59\,\Omega$ @ 134.2 kHz
$R_{Tx} = 15.1\,\Omega$ @ 123.2 kHz	$R_{Rx} = 42.75\,\Omega$ @ 123.2 kHz
$R_{Q_{Tx}} = 5\,\text{k}\Omega$	
$L_{Tx} = 443\,\mu\text{H}$	$L_{Rx} = 2.49\,\text{mH}$
Without considering $R_{Q_{Tx}}$:	$Q_{Rx} = 47.1$ @ 134.2 kHz
$Q_{Tx} \simeq \frac{\omega L_{Tx}}{R_{Tx}} = 23.5$ @ 134.2 kHz	$Q_{Rx} = 45.1$ @ 123.2 kHz
$\quad = 22.7$ @ 123.2 kHz	
The influence of $R_{Q_{Tx}}$ on the	
equivalent Q_{Tx} is addressed later	
C_{Tx} and C_{Rx} selected to resonate at 134.2 kHz	
$C_{Tx} = 3.17\,\text{nF}$	$C_{Rx} = 565\,\text{pF}$
$R_S = 47\,\text{k}\Omega$	$R_L = 2.2\,\text{M}\Omega$
	$C_L = 68\,\text{nF}$
V_S is a square wave with 5 V amplitude	V_T is modeled as a sinusoidal wave
Fundamental harmonic: $V_S = \frac{4}{\pi}5\,\text{V} \simeq 6.37\,\text{V}$	$V_T = 9\,\text{V}$
$R_{out_R} = 11.5\,\Omega$	
	$r_{Rx} = 1.35\,\text{cm}$ (outer)
Radius $r_{Tx} = 6\,\text{cm}$	Radius $\quad = 0.7\,\text{cm}$ (inner)
	$\quad = 1\,\text{cm}$ (average)
For numerical and simulation purpose, distance between coils is $D_{Tx\text{-}Rx} = 10\,\text{cm}$	

In this system, the maximum read distance can be limited by either the charging or the reading phase. The charging phase is considered successful if V_L in the tag surpasses 5 V before the end of the charging phase (based on ICAR guidelines [2]). The reading phase is possible if the amplitude of the signal received by the reader, V_{R_S}, is high enough to be decoded. The minimum acceptable V_{R_S} depends on the demodulator used and the noise level. In this example, a minimum required V_{R_S} of 10 mV was considered.

8.1.2 2-Coil RFID Link

8.1.2.1 Charging Phase

For the sake of simplicity, the $R_{Q_{Tx}}$ shown in Fig. 8.1a, added to decrease the reader coil quality factor and avoid Intersymbol Interference (ISI) which is further addressed in Sect. 8.1.3, is not going to be considered initially. Thus, the model is similar to the analysis presented in this book. At the end of this section, the influence of $R_{Q_{Tx}}$ is fully addressed by showing that it can be modeled as an increase in the coil Equivalent Series Resistance (ESR), R_{Tx}.

Let us start by calculating the PTE. Since efficiency depends on the distance between coils, $D_{Tx\text{-}Rx}$, we assume that the nominal $D_{Tx\text{-}Rx}$ is 10 cm for the numerical calculations and simulations.

The link efficiency, η_{Link}, for a 2-coil link, was analyzed in Sect. 2.1.3. In Table 8.2, the efficiency for this RFID link in the charging phase (Fig. 8.1a) is calculated clearly describing each step and referencing the equations used by their numbers in this book.

The received power, P_{MN} in Fig. 8.1a, was analyzed in Sect. 2.1.4 for different Tx-circuits, and it was summarized in Table 2.3. In this case, the reader (Tx) has a voltage source and a series resonant capacitor. Therefore its P_{MN} is

$$P_{MN} = \frac{V_S^2}{2R_{Tx}} \frac{Q_{Rx\text{-}L}}{Q_L} \frac{k_{Tx\text{-}Rx}^2 Q_{Tx} Q_{Rx\text{-}L}}{(k_{Tx\text{-}Rx}^2 Q_{Tx} Q_{Rx\text{-}L} + 1)^2} = 2.8 \text{ mW}. \tag{2.17}$$

To calculate P_{MN} from (2.17), note that R_{Tx} should be substituted by the reader coil ESR plus the driver output impedance ($R_{Tx} + R_{out_R}$ from Table 8.1). V_S can also be obtained from Table 8.1, and the rest of the parameters were calculated in Table 8.2.

Approximating the rectifier efficiency by one, i.e., $P_L = P_{MN}$, the load voltage, V_L, can be simply calculated as

$$\text{PDL} = P_L = P_{MN} = \frac{V_L^2}{R_L} \Rightarrow V_L = \sqrt{P_{MN} R_L} = 78 \text{ V}, \tag{8.1}$$

and the PTE is

$$\text{PTE} = \eta_{TOT} = \eta_{Link} \underbrace{\eta_{rect}}_{\simeq 1} = 0.4 \%. \tag{8.2}$$

Note that even with such a low PTE of 0.4%, the RFID link that is described here has a large voltage gain of $V_L/V_S = 78/6.37$ V/V. Therefore, in this system, the Rx-circuit should be able to tolerate and limit large voltages (tens of volts) across the Rx coil.

Table 8.2 Step-by-step calculation of the RFID link efficiency during charging phase. The system is presented in Fig. 8.1a and its component values in Table 8.1. The calculation script is available in the supplementary material, file Sec812_charging

From (2.13) $\Longrightarrow$ $\quad \eta_{Link} = \dfrac{Q_{Rx\text{-}L}}{Q_L} \dfrac{k_{Tx\text{-}Rx}^2 Q_{Tx} Q_{Rx\text{-}L}}{k_{Tx\text{-}Rx}^2 Q_{Tx} Q_{Rx\text{-}L} + 1}$ $= 0.4\%$

$$Q_{Tx} = \frac{\omega L_{Tx}}{R_{Tx} + R_{out_R}} = 13.6$$

The output resistance of the driver, R_{out_R}, is included as part of the the Tx coil ESR, $R_{Tx} + R_{out_R}$. All these parameters, $R_{Tx}, L_{Tx}, R_{out_R}$ and $\omega = 2\pi f$, can be found in Table 8.1.

$(3.16) \Rightarrow k_{Tx\text{-}Rx} = \dfrac{r_{Tx}^2 \cdot r_{Rx}^2}{\sqrt{r_{Tx} \cdot r_{Rx}} \left(\sqrt{D_{Tx\text{-}Rx}^2 + r_{Tx}^2}\right)^3} = 0.0093$

Deduced in Section 3.1.1, valid for circular, coplanar, perfectly aligned, coils with $r_{Tx} \geqslant r_{Rx}$. The distance between coils, $D_{Tx\text{-}Rx}$, and coils radius, r_{Tx} and r_{Rx}, are presented in Table 8.1.

$(2.11) \Rightarrow Q_{Rx\text{-}L} = \dfrac{Q_{Rx} Q_L}{Q_{Rx} + Q_L} = 43.2$

$Q_{Rx\text{-}L}$ is the Rx equivalent quality factor, defined in Section 2.1.3. Q_{Rx} is presented in Table 8.1 and Q_L is calculated below.

$(2.4) \Rightarrow Q_L = \dfrac{\omega L_{Rx}}{Re\{Z_{MN}\}} = 523.9$

Q_L is the load quality factor, defined in Section 2.1.3. $\omega = 2\pi f$ and L_{Tx} can be obtained from Table 8.1. The $Re\{Z_{MN}\}$ is calculated below.

$(5.28) \Rightarrow Re\{Z_{MN}\} = \dfrac{1}{R_{rect}(\omega C_{Rx})^2} = 4\,\Omega$

Z_{MN} is the input impedance of the matching network, which in this case was implemented by a parallel capacitor, Fig. 8.1a. The input impedances of various matching networks were studied in Section 5.6, and were summarized in Table 5.4. $Re\{Z_{MN}\}$ is the real part of Z_{MN}. From Table 8.1, $\omega = 2\pi f$ and C_{Rx} can be obtained. The R_{rect} is calculated below.

$(6.3) \Rightarrow R_{rect} = \dfrac{R_L \cdot \eta_{rect}}{2 G_{rect}^2} \simeq \dfrac{R_L}{2} = 1.1\,\text{M}\Omega$

The input resistance of a rectifier, R_{rect}, was studied in Section 6.3. In this case, a DC-DC converter is not included in the model (Fig. 8.1a), therefore, the rectifier is loaded with R_L instead of being loaded with the input resistance of a DC-DC converter, $R_{DC\text{-}DC}$, as in Section 6.3. Since a simple half-wave rectifier is being used here, and neglecting the diode voltage drop, we will approximate both η_{rect} and G_{rect} by one. The load, R_L, can be obtained from Table 8.1.

The calculations presented in this section can be found in the supplementary material, file Sec812_charging. This circuit (Fig. 8.1a) was also simulated using LTspice. The LTspice schematic is shown in Fig. 8.2, and it can also be found in the supplementary material, file Sec812_charging_spice, for the readers to be able to rerun simulations, change design parameters, and see the results. After running the simulation, click on *View → SPICE Error Log* (or use the Ctrl-L shortcut) to view the post-processing results which include the PTE and PDL, among other relevant parameters.

The simulated results are in good agreement with the previously calculated values, as shown in Table 8.3. The input resistance of the rectifier was deduced from the simulation by using the first harmonic of the rectifier input current and the first harmonic of the rectifier input voltage, as detailed in Table 8.4. In this case, the simulated rectifier input capacitance is 859 fF, which can be neglected as it is in parallel with $C_{Rx} = 565$ pF. The main reason for the slight differences between calculations and simulations in this case is the simplified model used for the rectifier. If the circuit shown in Fig. 8.3 is simulated (supplementary material, file Sec812_charging_spice_simp) where the rectifier was substituted by its calculated input resistance, the results obtained are almost identical to the calculation.

So far, the $R_{Q_{Tx}} = 5$ kΩ (Fig. 8.1a) has not been considered. As mentioned in Sect. 4.3.2, this resistance is included to reduce the reader coil quality factor and avoid ISI during the reading phase, which is further addressed in Sect. 8.1.3. In order to address its effect on the charging phase, let us calculate the impedance

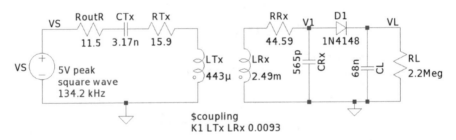

Fig. 8.2 LTspice schematic of 2-coil RFID inductive link in the charging phase. The LTspice schematic is available in the supplementary material, file Sec812_charging_spice

Table 8.3 Comparison between calculated and simulated results of 2-coil RFID inductive link in the charging phase. The calculation script and the LTspice schematic are available in the supplementary material, files Sec812_charging and Sec812_charging_spice, respectively

Result	Calculated	Simulated
PTE	0.40%	0.39%
PDL	2.80 mW	2.73 mW
V_L	78.5 V	77.5 V
R_{rect}	1.1 MΩ	1.1 MΩ

Table 8.4 Deduction of R_{rect} from simulation

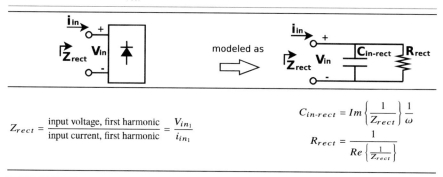

$Z_{rect} = \dfrac{\text{input voltage, first harmonic}}{\text{input current, first harmonic}} = \dfrac{V_{in_1}}{i_{in_1}}$	$C_{in\text{-}rect} = Im\left\{\dfrac{1}{Z_{rect}}\right\}\dfrac{1}{\omega}$ $R_{rect} = \dfrac{1}{Re\left\{\dfrac{1}{Z_{rect}}\right\}}$

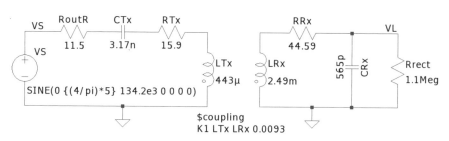

Fig. 8.3 LTspice simplified schematic of 2-coil RFID inductive link in the charging phase. The LTspice schematic is available in the supplementary material, file Sec812_charging_spice_simp

Fig. 8.4 Effect of $R_{Q_{Tx}}$ on the Tx coil quality factor

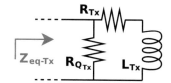

seen through the Tx coil, $Z_{eq\text{-}Tx}$, defined in Fig. 8.4. $Z_{eq\text{-}Tx}$ is calculated in (8.3), assuming that $R_{Q_{Tx}} \gg R_{Tx}$ and $R_{Q_{Tx}} \gg \omega L_{Tx}$, which is valid in actual cases as $R_{Q_{Tx}}$ is added in parallel to reduce the quality factor, but the inductor should dominate (have a lower impedance). In this case, this hypothesis is valid as

$$\underbrace{R_{Q_{Tx}}}_{5\,k\Omega} \gg \underbrace{\omega L_{Tx}}_{373.5\,\Omega} \gg \underbrace{R_{Tx}}_{15.9\,\Omega}.$$

$$Z_{eq\text{-}Tx} = \frac{(R_{Tx} + j\omega L_{Tx})R_{QTx}}{R_{Tx} + j\omega L_{Tx} + R_{QTx}} \xrightarrow{R_{QTx} \gg R_{Tx}}$$

$$= \frac{(R_{Tx} + j\omega L_{Tx})R_{QTx}}{R_{QTx} + j\omega L_{Tx}} \cdot \frac{R_{QTx} - j\omega L_{Tx}}{R_{QTx} - j\omega L_{Tx}}$$

$$= \frac{(R_{Tx} + j\omega L_{Tx})R_{QTx}(R_{QTx} - j\omega L_{Tx})}{R_{QTx}^2 + (\omega L_{Tx})^2}$$

$$= R_{QTx}\frac{R_{Tx}R_{QTx} + (\omega L_{Tx})^2 + j\omega L_{Tx}(R_{QTx} - R_{Tx})}{R_{QTx}^2 + (\omega L_{Tx})^2} \xrightarrow[R_{QTx} \gg R_{Tx}]{R_{QTx} \gg \omega L_{Tx}}$$

$$= R_{QTx}\frac{R_{Tx}R_{QTx} + (\omega L_{Tx})^2 + j\omega L_{Tx}R_{QTx}}{R_{QTx}^2}$$

$$= R_{Tx} + \frac{(\omega L_{Tx})^2}{R_{QTx}} + j\omega L_{Tx}$$

$$(8.3)$$

As can be seen from (8.3), R_{QTx} can be modeled as an increase in the Tx coil ESR, i.e., as a reduction in its quality factor, Q_{Tx}. Therefore, including $R_{QTx} = 5\ k\Omega$ and also considering the driver output impedance as part of the Tx coil ESR,

$$Q_{Tx} = \frac{\omega L_{Tx}}{R_{Tx} + R_{out_R} + \frac{(\omega L_{Tx})^2}{R_{QTx}}} = 6.8.$$

To consider R_{QTx} in the previously used script, file Sec812_charging (supplementary material), uncomment the lines 47 and 53. The LTspice including R_{QTx} is shown in Fig. 8.5 and can be found in the supplementary material, file Sec812_charging_spice_RQTX.

The calculated and simulated performance of the link in this case, including R_{QTx}, is summarized in Table 8.5. As can be seen, comparing Tables 8.5 and 8.3, the performance of the link during the charging phase worsened due to R_{QTx}. However,

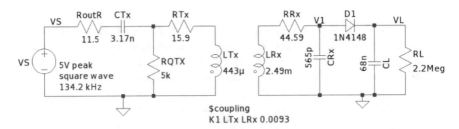

Fig. 8.5 LTspice schematic of 2-coil RFID inductive link in the charging phase including R_{QTx}. The LTspice schematic is available in the supplementary material, file Sec812_charging_spice_RQTX

Table 8.5 Comparison between calculated and simulated results of 2-coil RFID inductive link in the charging phase including $R_{Q_{Tx}}$ to avoid ISI. The calculation script and the LTspice schematic are available in the supplementary material, files Sec812_charging and Sec812_charging_spice_RQTX, respectively

Result	Calculated	Simulated
PTE	0.20%	0.19%
PDL	721 μW	691 μW
V_L	40 V	39 V
R_{rect}	1.1 MΩ	1.1 MΩ

as further discussed in Sect. 8.1.3, to improve system reliability, it is necessary to have low-quality factors in order to avoid ISI.

This system, including $R_{Q_{Tx}}$, corresponds to the actual case of a low-cost RFID reader designed by Universidad de la República, Uruguay. The read distance of this reader and the actual tag (RI-INL-R9QM) was presented in [4] and summarized in Sect. 4.3.2. In measurements, the maximum read distance of the actual tag was 16 cm, and it was proved that the tag fails to be read correctly due to the charging phase by not receiving enough power for distances larger than 16 cm, as shown in Sect. 4.3.2. As explained in Sect. 8.1.1, the read distance is considered successful in this model if V_L is larger than 5 V at the end of the charging phase. Using the script file Sec812_charging (supplementary material), the maximum read distance, i.e., distance for which $V_L \simeq 5$ V, is $\simeq 25$ cm. In LTspice simulations using file Sec812_charging_spice (supplementary material) with the coupling coefficient calculated from (3.16), the distance for which $V_L \simeq 5$ V is $\simeq 21$ cm. In [4], the model was improved using the coupling coefficient simulated with a field solver software (CST), and the model predicts a maximum read distance of $\simeq 20$ cm.

8.1.2.2 Reading Phase

The procedure to calculate the PTE and PDL in the reading phase (Fig. 8.1b) is similar to the charging phase, while the priorities are different. During the reading phase, the tag, which was previously referred to as the Rx, becomes the Tx and transmits its ID using the power stored during the charging phase. On the other side, the reader becomes the Rx, receiving and decoding the tag ID. Therefore, to calculate the efficiency using (2.13), in this case Q_{Tx} and Q_{Rx} should be substituted by the tag and reader quality factors, respectively. Additionally, during this phase, the load resistance is the parallel combination of $R_{Q_{Tx}}$ and R_S, as shown in Fig. 8.1b.

Note that during the reading phase, the model used for the tag, which is based on ICAR guidelines, is not resonating as it has no resonant capacitor. Additionally, since each bit is represented by a different frequency, the Rx (i.e., the reader) is only resonating in the case of a bit "0," which is represented by the same frequency as the carrier in the charging phase. The Rx is considered detuned when it is receiving the bit "1," which is modulated with a lower frequency. Non-resonant WPT links were addressed in Sects. 2.1.6 and 2.1.7.

From the calculated PDL, which in this case is the power delivered to the parallel combination of $R_{Q_{Tx}}$ and R_S, the voltage amplitude across these resistances, V_{R_S}, can be calculated. Note that in this case, where the goal is to decode the tag's ID from V_{R_S} instead of transferring power, the amplitude of V_{R_S} is more relevant than the values of PTE and PDL. At $D_{Tx\text{-}Rx} = 10\,$cm, $V_{R_S} = 696\,$mV, and $V_{R_S} = 215\,$mV for the "0" and "1" bits, respectively. As mentioned in Sect. 8.1.1, a minimum required V_{R_S} of 10 mV was considered to be able to decode the tag ID, which is reached by the bit "1" at $D_{Tx\text{-}Rx} = 32\,$cm. Therefore, the reading phase would be considered successful up to $D_{Tx\text{-}Rx} = 32\,$cm; however, at such a distance, the charging phase fails, as analyzed in the previous section.

The calculation script and LTspice schematic for this case can be found in the supplementary material, files Sec812_reading, Sec812_reading_low_spice, and Sec812_reading_high_spice, respectively.

It should be mentioned that the discussion presented in this section and the calculation scripts provided are not considering the transient response. The reading phase can fail not only due to a low amplitude in V_{R_S} but also due to ISI, as previously mentioned and is further analyzed for the case of a 3-coil link in the next section.

8.1.3 3-Coil RFID Link

As discussed in Sect. 2.2, PTE and PDL can be improved by using additional resonators. In this section, a resonator is added between the reader and tag to improve the RFID read distance. The resonator parameters are summarized in Table 8.6, and its design was presented in [4]. In this section, we address the effect of this resonator on the RFID link. First, in Sect. 8.1.3.1, the improvements in the charging phase are calculated. Then in Sect. 8.1.3.2, the resonator effect on the reading phase is addressed.

8.1.3.1 Charging Phase

The lumped circuit model for the RFID 3-coil link during the charging phase is shown in Fig. 8.6. The link efficiency for this case was analyzed in Sect. 2.2.1:

$$\eta_{Link} = \frac{Q_{Rx\text{-}L}}{Q_L} \frac{k_{A\text{-}Rx}{}^2 Q_A Q_{Rx\text{-}L}}{(k_{A\text{-}Rx}{}^2 Q_A Q_{Rx\text{-}L} + 1)} \frac{k_{Tx\text{-}A}{}^2 Q_{Tx} Q_A}{(k_{Tx\text{-}A}{}^2 Q_{Tx} Q_A + k_{A\text{-}Rx}{}^2 Q_A Q_{Rx\text{-}L} + 1)}. \tag{2.32}$$

To calculate η_{Link} from (2.32), note that Q_A can be obtained from Table 8.6, $k_{Tx\text{-}A}$ and $k_{A\text{-}Rx}$ can be calculated using (3.16) as was done in the 2-coil case, and all the other parameters in (2.32) were calculated in Table 8.2.

Table 8.6 Additional resonator parameter. Designed in [4]

Parameter	Value
L_A	648 μH
R_A	13.3 Ω @ 134.2 kHz
	11.0 Ω @ 123.2 kHz
Q_A	41 @ 134.2 kHz
	45.6 @ 123.2 kHz
Radius r_A	13 cm
For numerical and simulation purpose, distance between coils is:	
Between reader and additional resonator	$D_{Tx\text{-}A} = 30$ cm
Between additional resonator and tag	$D_{A\text{-}Rx} = 5$ cm

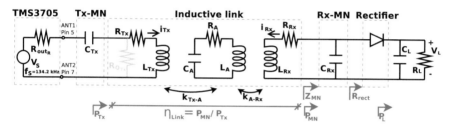

Fig. 8.6 Lumped circuit model of 3-coil RFID link in the charging phase

The PDL for a 3-coil link was also deduced in Sect. 2.2:

$$P_{MN} = \eta_{Link} P_{Tx} = \frac{V_S^2}{2R_{Tx}} \frac{Q_{Rx\text{-}L}}{Q_L} \frac{(k_{A\text{-}Rx}^2 Q_A Q_{Rx\text{-}L})(k_{Tx\text{-}A}^2 Q_{Tx} Q_A)}{(k_{Tx\text{-}A}^2 Q_{Tx} Q_A + k_{A\text{-}Rx}^2 Q_A Q_{Rx\text{-}L} + 1)^2}.$$
(2.34)

Just like in the 2-coil case, to consider the driver output resistance, the R_{Tx} in (2.34) should be substituted by the reader coil ESR plus the driver output resistance ($R_{Tx} + R_{out_R}$ from Table 8.1). Then, approximating the rectifier efficiency by one, i.e., $P_L = P_{MN}$, the load voltage, V_L, can be simply calculated as

$$\text{PDL} = P_L = P_{MN} = \frac{V_L^2}{R_L} \Rightarrow V_L = \sqrt{P_{MN} R_L}.$$

The calculation script and LTspice schematic for this 3-coil RFID link can be found in the supplementary material, files Sec813 and Sec813_charging_spice, respectively.

For instance, if $R_{Q_{Tx}}$ is not included

$$
\begin{array}{llll}
\text{Calculation} & \Longrightarrow & \text{PTE} = 0.36\% & ; & V_L = 71 \text{ V.} \\
\text{Simulation} & \Longrightarrow & \text{PTE} = 0.34\% & ; & V_L = 69 \text{ V.}
\end{array}
$$
(8.4)

The results obtained in this 3-coil case, with a total distance between reader and tag of $D_{Tx\text{-}Rx} = D_{Tx\text{-}A} + D_{A\text{-}Rx} = 35$ cm, are similar to the ones obtained with the 2-coil link at $D_{Tx\text{-}Rx} = 10$ cm (see Table 8.3 and (8.4)). This proves the improvements in the charging phase in terms of viable charging distance by a factor of $3.5\times$, thanks to the resonator. However, that is not the entire story, and to improve the read distance, this extra resonator should also extend the range during the reading phase, which is studied in the next section.

8.1.3.2   Reading Phase

During the reading phase, the reader has to decode the tag response which is modulated in frequency. The bit "0" is represented by the same frequency as the carrier in the charging phase, 134.2 kHz, while the bit "1" is represented by a lower frequency of 123.2 kHz. Each bit is expected to be differently affected by the resonator, which was designed to resonate at the carrier frequency during the charging phase (134.2 kHz).

Figure 8.7 shows the reading phase simulation results in LTspice. The tag is transmitting a stream of random bits, which are decoded by the reader using a Phase-Locked Loop (PLL).

In Fig. 8.8, the simulated, transmitted, and received bits are presented. As can be seen, the reader is not being able to decode the tag ID. This is because, after the transmission of a bit, the resonant system remains oscillating enough time to interfere with the following bit, which is known as Intersymbol Interference (ISI). The greater the coil quality factor, the more time each bit requires before dying out. This is aggravated by the large difference in the amplitude of each bit due to resonance. The bit "0," which is transmitted at the resonant frequency, is received (V_{Rs} in Fig. 8.7) with an amplitude that is around 18 times larger than the bit "1," which is transmitted out of resonance. Therefore, when the transmission of a bit "1" starts after the transmission of a bit "0," the relatively lower received amplitude of

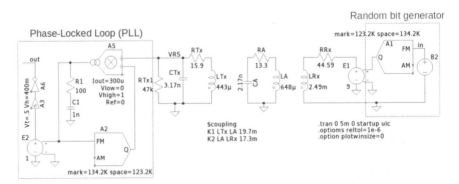

Fig. 8.7 LTspice schematic of 3-coil RFID inductive link in the reading phase. The LTspice schematic is available in the supplementary material, file Sec813_reading_not_working_spice

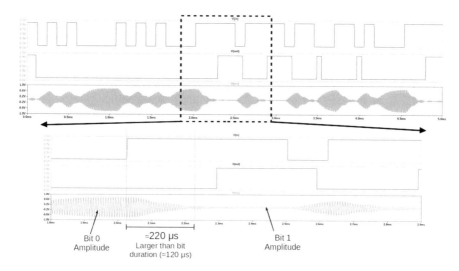

Fig. 8.8 LTspice simulation results of 3-coil RFID link in the reading phase. The schematic is shown in Fig. 8.7. The LTspice schematic is available in the supplementary material, file Sec813_reading_not_working_spice

the bit "1" follows the residual of the relatively larger amplitude of the previous bit "0," causing even more ISI. V_{Rs} is also presented in Fig. 8.8. First, note the large difference in V_{Rs} amplitude for each bit. Then, note that when the transmission of a bit "0" ends, the ISI remains active in the link for a time that is larger than the duration of each bit.

To reduce ISI in the reading phase, the reader is detuned by adding $C_{Tx_{extra}}$, and the resonator quality factor is reduced by adding R_{Q_A}, as shown in Fig. 8.9. The capacitor $C_{Tx_{extra}} = 390$ pF sets the reader resonant frequency at a value that is between the bits "0" and "1" frequencies:

$$\overbrace{123.2 \text{ kHz}}^{\text{bit '0'}} < \overbrace{126.7 \text{ kHz}}^{\text{reader resonance}} < \overbrace{134.2 \text{ kHz}}^{\text{bit '1'}}.$$

This tends to match the "0" and "1" bit amplitudes. The R_{Q_A} reduces the resonator quality factor which was larger in the link, to accelerate the rate of reduction in the amplitude of both bits. By reducing the resonator quality factor, the difference between bit amplitudes is also reduced. R_{Q_A} has the same effect in the resonator as $R_{Q_{Tx}}$ in the reader (8.3); thus, $Q_A = \frac{\omega L_A}{R_A + (\omega L_A)^2 / R_{Q_A}}$. In this 3-coil link, $R_{Q_{Tx}}$ can also be added to reduce the bit decay time even more, but as shown in simulation results, it is not necessary. The simulation results of this circuit (Fig. 8.9), including $C_{Tx_{extra}}$ and R_{Q_A}, are shown in Fig. 8.10. As can be seen, the tag response is correctly decoded now. This LTspice schematic can be found in the supplementary material, file Sec813_reading_spice; it is interesting to simulate the circuit and verify that both $C_{Tx_{extra}}$ and R_{Q_A} are needed to make the reading phase possible.

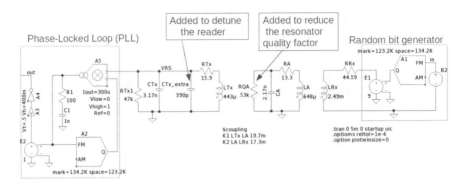

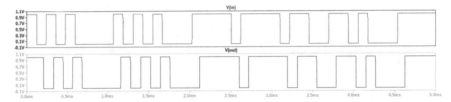

Fig. 8.9 LTspice schematic of 3-coil RFID inductive link in the reading phase including $C_{Tx_{extra}}$ and R_{Q_A} to avoid ISI. The LTspice schematic is available in the supplementary material, file Sec813_reading_spice

Fig. 8.10 LTspice simulation results of 3-coil RFID link in the reading phase including $C_{Tx_{extra}}$ and R_{Q_A} to avoid ISI. The schematic is shown in Fig. 8.9. The LTspice schematic is available in the supplementary material, file Sec813_reading_spice

Table 8.7 Comparison between calculated and simulated results of 3-coil RFID inductive link in the charging phase including $C_{Tx_{extra}}$ and R_{Q_A}. The calculation script and LTspice schematic are available in the supplementary material, files Sec813 and Sec813_charging2_spice, respectively

Result	Calculated	Simulated
PTE	0.22%	0.21%
PDL	535 μW	486 μW
V_L	34 V	33 V

The inclusion of both $C_{Tx_{extra}}$ and R_{Q_A} deteriorates the charging phase. To include these two components in the script file Sec813 (supplemenary material), uncomment lines 36 and 66. The LTspice schematic including this component in the charging phase can be found in the supplementary material, file Sec813_charging2_spice. The results obtained in this case are summarized in Table 8.7.

Comparing Table 8.7 with the values obtained in (8.4) without $C_{Tx_{extra}}$ and R_{Q_A}, the performance reduction in the charging phase is evident. However, this is necessary for the tag to be operational in both phases, charging and reading.

In [4], the read distance of this RFID link was increased from 16 to 43 cm by adding the extra resonator, proving the advantages of using resonators in such RFID links.

8.2 Introduction to WPT Links for Visual Prosthesis

A visual prosthesis is a real-time interface between the nervous system and the external world, through an artificial image sensor, which tries to mimic the human visual system. Nowadays, these prostheses are able to provide a rudimentary sense of vision, but they have not yet been widely utilized clinically in the blind due to limited benefits that they offer compared to the immense complexity of the human visual system [5].

A visual prosthesis is composed of an external system and an implant. In the external system, a video camera captures images at a certain rate, which are processed and wirelessly transmitted to the implant. Depending on the type of visual prosthesis, the implant stimulates the user's visual system at a certain anatomical position, e.g., in the retina, optic nerve, or visual cortex [6].

The implanted part of a visual prosthesis is a size-constrained device whose power requirements cannot be met by a battery. WPT has demonstrated to be a suitable method to power these devices [7–10]. In that case, the Tx coil is placed in the external driver, which can be part of the glasses or hidden behind the ears, at 1 inch or closer to the Rx implanted coil [9].

Further information regarding visual prostheses can be found in references, such as [6, 11]. In this section, we address a WPT system with coil sizes and load power requirements that are appropriate for retinal or cortical visual prostheses. We particularly focus on the design of the Rx matching network and Optimum Operating Point (OOP) tracking, thus reviewing concepts of Chaps. 2, 5, and 6.

First, in Sect. 8.2.1, the example of a WPT link for visual prostheses is presented. Then, in Sect. 8.2.2, the Rx matching network is addressed, comparing the series versus parallel compensation in this actual case. The loading of a retinal implant would be quite different when the patient is in a dark room, versus looking at a bright image; therefore, in Sect. 8.2.3, the OOP tracking under load variations is addressed.

8.2.1 WPT Link for Visual Prostheses

The coils used in this link are described in Table 8.8. In their design, presented in [8], the theoretical models introduced in Sect. 3.1.1 for Printed Spiral Coils (PSC) were

Table 8.8 Coils for visual prosthesis WPT link

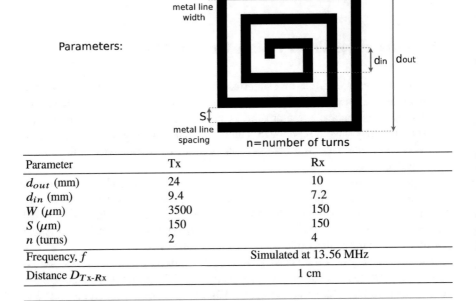

Parameter	Tx	Rx
d_{out} (mm)	24	10
d_{in} (mm)	9.4	7.2
W (μm)	3500	150
S (μm)	150	150
n (turns)	2	4
Frequency, f		Simulated at 13.56 MHz
Distance $D_{Tx\text{-}Rx}$		1 cm

used. However, as mentioned in Chap. 3, although theoretical models can be used for preliminary estimations and theoretical optimization of the coil geometry, computer-aided design tools generally known as field solvers (see Table 2.4) are usually required for more accurate and complex designs. Therefore, the coil equivalent lumped model, presented in Table 8.9, was obtained from simulations (HFSS, ANSYS), including muscle tissue surrounding the Rx coil. This link, considering the size and separation of the coils, is reasonable for a retinal or cortical visual prosthesis.

The parallel resistance, R_P, shown in Table 8.9, can be modeled in series with the coil as was deduced in (8.3) for the previous RFID example. That allows us to match the coil model presented in Table 8.9 to the analysis presented in the book. Additionally, the coils should be used at a frequency that is far enough from their Self-Resonant Frequency (SRF), preferably a decade lower, because at frequencies greater than the coil SRF its parasitic capacitance dominates over its inductance and the inductive link performance is rapidly degraded. In this case, the carrier frequency is 13.56 MHz, which is more than a decade below the coil SRF (see Table 8.9). Therefore, the parasitic capacitors, C_P can be neglected. This equivalent simplified model is presented in Table 8.10.

The power required by the implant, P_L, depends on the specific design and operation conditions. In [6], P_L was estimated to be 36 mW, while in [9] it was estimated to be 45 mW. In this example, we approximate the maximum power required by the load to be about 50 mW. The voltage required also depends on the

Table 8.9 Tx and Rx coil lumped equivalent circuits obtained from simulation using HFSS, ANSYS

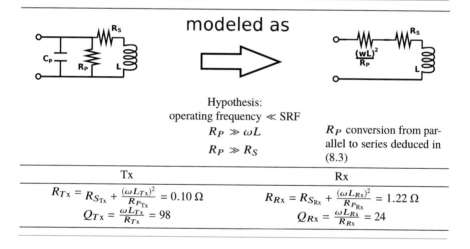

Coil lumped equivalent circuit	
Tx	Rx
$L_{Tx} = 120$ nH	$L_{Rx} = 340$ nH
$R_{S_{Tx}} = 60$ mΩ	$R_{S_{Rx}} = 720$ mΩ
$R_{P_{Tx}} = 2.36$ kΩ	$R_{P_{Rx}} = 1.68$ kΩ
$C_{P_{Tx}} = 7.72$ pF	$C_{P_{Rx}} = 0.77$ pF
$SRF = 165$ MHz	$SRF = 302$ MHz
At $D_{Tx\text{-}Rx} = 1$ cm $\Longrightarrow k_{Tx\text{-}Rx} = 0.0301$	

Table 8.10 Tx and Rx coil simplified models

modeled as

Hypothesis:
operating frequency ≪ SRF
$$R_P \gg \omega L$$
$$R_P \gg R_S$$

R_P conversion from parallel to series deduced in (8.3)

Tx	Rx
$R_{Tx} = R_{S_{Tx}} + \dfrac{(\omega L_{Tx})^2}{R_{P_{Tx}}} = 0.10\ \Omega$	$R_{Rx} = R_{S_{Rx}} + \dfrac{(\omega L_{Rx})^2}{R_{P_{Rx}}} = 1.22\ \Omega$
$Q_{Tx} = \dfrac{\omega L_{Tx}}{R_{Tx}} = 98$	$Q_{Rx} = \dfrac{\omega L_{Rx}}{R_{Rx}} = 24$

circuit implementation. We assume that a constant DC voltage of 5 V is required, which is enough for the electrical stimulation [5]. Therefore, the load resistance is estimated as 500 Ω = (5 V)^2/50 mW. To provide the constant DC voltage, a rectifier and a voltage regulation feedback loop are required. In this example, we will use the active cross-connected rectifier presented in Fig. 6.5d and analyzed in Sect. 6.3. The voltage regulation is implemented by adjusting the Tx voltage, i.e., (A)-type feedback of Fig. 7.1, introduced in Sect. 7.1.

The schematic of the entire WPT link is depicted in Fig. 8.11, using the simplified coil models presented in Table 8.10. The design of the Rx matching network is addressed in Sect. 8.2.2. Then, in Sect. 8.2.3, the use of the reconfigurable active rectifier presented in Sect. 6.3.2 (Fig. 6.10) to track the OOP is discussed.

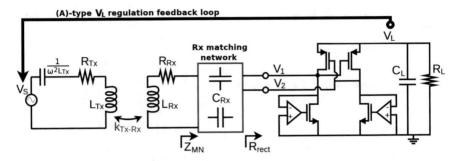

Fig. 8.11 Example of WPT link for visual prosthesis analyzed in this section

8.2.2   Rx Matching Network Design: Series Versus Parallel

In this section, the series and parallel Rx matching networks in this visual prosthesis example are compared using the tools developed in the book to justify this design decision. The series and parallel Rx matching networks were initially compared in Sect. 2.1.6. This discussion was broadened in Sect. 5.6, where it was proved that the Rx matching network should be selected to operate as close to the Optimum Operating Point (OOP) as possible. As discussed in Chap. 5, two optimum values exist for the Rx-circuit input impedance, Z_{MN}. One of them maximizes the PTE, $Z_{MN_{opt-\eta}}$, achieving the Maximum Efficiency Point (MEP), while the other maximizes the PDL, $Z_{MN_{opt-P_{MN}}}$, achieving the Maximum Power Point (MPP). The OOP was used to refer to either MEP or MPP depending on the desired link optimization paradigm, which is in turn linked to the application. In both cases, the imaginary part of this optimum Z_{MN} is the one that achieves resonance in the Rx, $Im\{Z_{MN_{opt-\eta}}\} = Im\{Z_{MN_{opt-P_{MN}}}\} = -\omega L_{Rx}$. The optimum real parts were deduced in Sects. 5.2 and 5.3, which are recapitulated and evaluated for this example in the following:

$$(5.4) \Rightarrow Re\{Z_{MN_{opt-\eta}}\} = R_{Rx}\sqrt{k_{Tx\text{-}Rx}^2 Q_{Tx} Q_{Rx} + 1} = 2.15\ \Omega$$

$$(5.9) \Rightarrow Re\{Z_{MN_{opt-P_{MN}}}\} = R_{Rx}(k_{Tx\text{-}Rx}^2 Q_{Tx} Q_{Rx} + 1) = 3.79\ \Omega \tag{8.5}$$

Note that $k_{Tx\text{-}Rx}$ is presented in Table 8.9, while Q_{Tx}, Q_{Rx}, and R_{Rx} are presented in Table 8.10.

In Table 8.11, the Z_{MN} obtained with a parallel and a series matching network are deduced for this example to see which one sets a Z_{MN} that is closer to the OOP.

As can be seen from Table 8.11, the parallel matching network sets the real part of Z_{MN} closer to both $Re\{Z_{MN_{opt-\eta}}\}$ (MEP) and $Re\{Z_{MN_{opt-P_{MN}}}\}$ (MPP), which were deduced in (8.5).

The choice between the MEP and the MPP was discussed in Sect. 5.4. The MEP ensures an efficient use of power and operating at this point is important to achieve low-power dissipation, low-temperature operation, and extended Tx lifetime when it is powered from a battery. These could be the highest priorities in an AIMD application like this. Therefore, the system should operate at the MEP as long as the Tx-circuit is capable of achieving the desired P_L in those conditions. In this example, P_L is held at 50 mW, thanks to the (A)-type feedback loop shown in Fig. 8.11.

Similar to the RFID example presented in Sect. 8.1, the step-by-step deduction of the link efficiency for this visual prosthesis is presented in Table 8.12 and evaluated for both parallel and series matching networks.

The script with all the theoretical calculations presented in this section can be found in the supplementary material, file Sec822.

Table 8.11 Calculation of Z_{MN} for series and parallel matching network

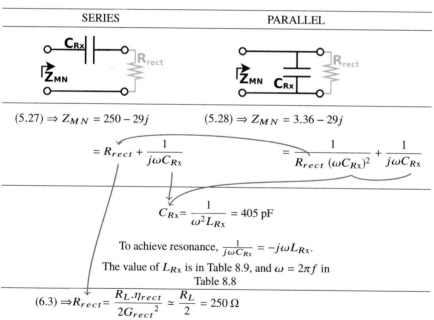

SERIES	PARALLEL

$(5.27) \Rightarrow Z_{MN} = 250 - 29j$ $(5.28) \Rightarrow Z_{MN} = 3.36 - 29j$

$$= R_{rect} + \frac{1}{j\omega C_{Rx}}$$

$$= \frac{1}{R_{rect}(\omega C_{Rx})^2} + \frac{1}{j\omega C_{Rx}}$$

$$C_{Rx} = \frac{1}{\omega^2 L_{Rx}} = 405 \text{ pF}$$

To achieve resonance, $\frac{1}{j\omega C_{Rx}} = -j\omega L_{Rx}$.

The value of L_{Rx} is in Table 8.9, and $\omega = 2\pi f$ in Table 8.8

$$(6.3) \Rightarrow R_{rect} = \frac{R_L \cdot \eta_{rect}}{2G_{rect}^2} \simeq \frac{R_L}{2} = 250 \ \Omega$$

The input resistance of a rectifier, R_{rect}, was studied in Section 6.3. In this case, a DC-DC converter is not included in the model (Fig. 8.11), therefore, the rectifier is loaded with R_L instead of being loaded with the input resistance of a DC-DC converter, $R_{DC\text{-}DC}$, as in Section 6.3. We have neglected the voltage drop and power losses in the rectifier, i.e. $\eta_{rect} = 1$ and $G_{rect} = 1$. The load, R_L, was estimated to 500 Ω in Section 8.2.1.

Table 8.12 Step-by-step calculation of the visual prosthesis link efficiency. The system is presented in Fig. 8.11 and its component values in Tables 8.8, 8.9, and 8.10. The calculation script is available in the supplementary material, file Sec·822

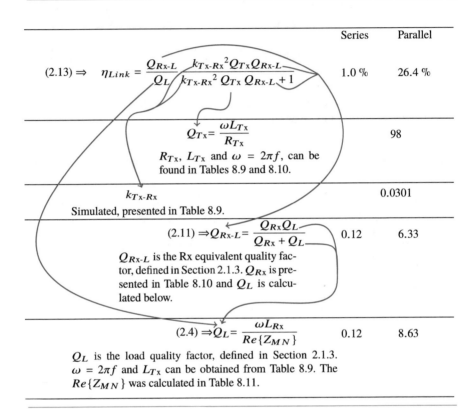

	Series	Parallel
$(2.13) \Rightarrow \quad \eta_{Link} = \dfrac{Q_{Rx\text{-}L}}{Q_L} \dfrac{k_{Tx\text{-}Rx}^2 Q_{Tx} Q_{Rx\text{-}L}}{k_{Tx\text{-}Rx}^2 Q_{Tx} Q_{Rx\text{-}L} + 1}$	1.0 %	26.4 %
$Q_{Tx} = \dfrac{\omega L_{Tx}}{R_{Tx}}$ R_{Tx}, L_{Tx} and $\omega = 2\pi f$, can be found in Tables 8.9 and 8.10.		98
$k_{Tx\text{-}Rx}$ Simulated, presented in Table 8.9.		0.0301
$(2.11) \Rightarrow Q_{Rx\text{-}L} = \dfrac{Q_{Rx} Q_L}{Q_{Rx} + Q_L}$ $Q_{Rx\text{-}L}$ is the Rx equivalent quality factor, defined in Section 2.1.3. Q_{Rx} is presented in Table 8.10 and Q_L is calculated below.	0.12	6.33
$(2.4) \Rightarrow Q_L = \dfrac{\omega L_{Rx}}{Re\{Z_{MN}\}}$	0.12	8.63

Q_L is the load quality factor, defined in Section 2.1.3. $\omega = 2\pi f$ and L_{Tx} can be obtained from Table 8.9. The $Re\{Z_{MN}\}$ was calculated in Table 8.11.

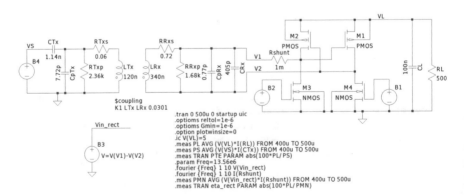

Fig. 8.12 LTspice schematic of the visual prosthesis with parallel matching network. The LTspice schematic is available in the supplementary material, file Sec822_parallel

Table 8.13 Calculated and simulated PTE for the visual prosthesis WPT link. LTspice schematic for the parallel case shown in Fig. 8.12. The calculation script and LTspice schematics are available in the supplementary material, files Sec822, Sec822_parallel, and Sec822_series

	Series		Parallel	
	Calculated	Simulated	Calculated	Simulated
PTE	1.0 %	0.7 %	26.4 %	25.0 %

This system was also simulated in LTspice, with the parallel and the series matching networks. The case with a parallel matching network is shown in Fig. 8.12. The schematics can be found in the supplementary material, files Sec822_parallel and Sec822_series. Note that the LTspice model includes not only the complete model of the coils and the active rectifier but also the (A)-type feedback loop. The feedback loop was implemented using the LTspice behavioral voltage source.

The simulated PTE for the series and parallel matching network are compared to the theoretical model in Table 8.13.

In summary, the parallel matching network achieves higher PTE as it sets a $Re\{Z_{MN}\}$ that is closer to the optimum value, $Re\{Z_{MN_{opt-\eta}}\}$, than the series matching network. The choice between series and parallel matching network was explained in this example.

However, as discussed in Chap. 6, the OOP is affected by variations in the coupling coefficient and load power consumption. In the next section, we discuss how to address the latter, OOP tracking under load variations.

8.2.3   Tracking OOP Under Load Variations

In this section, we use the parallel matching network as it was proved to achieve better performance than the series one.

As mentioned, the loading of a retinal implant would be quite different when the patient is in a dark room, versus looking at a bright image. Let us assume that the power consumed by the visual prosthesis drops from 50 to 10 mW due to injecting smaller currents in the tissue. In that case, the new value of the load is $R_L = \frac{(5\text{ V})^2}{10\text{ mW}} = 2500\ \Omega$.

If we repeat the calculations presented in the previous section using $R_L = 2500\ \Omega$, with the parallel matching network, we obtain $Re\{Z_{MN}\} = 0.67\ \Omega$. The $Re\{Z_{MN}\}$ with $R_L = 2500\ \Omega$ is further away from the optimal value $(Re\{Z_{MN_{opt-\eta}}\} = 2.15\ \Omega)$ than the $Re\{Z_{MN}\}$ with $R_L = 500\ \Omega$ (previously analyzed case). Therefore, the calculated efficiency with $R_L = 2500\ \Omega$ is 20.5 %, around 6 % less than with $R_L = 500\ \Omega$.

To maintain high efficiency, i.e., operate close to the MEP under load variations, some of the active circuits presented in Chap. 6 can be used. In this example, we use the reconfigurable active rectifier introduced in Sect. 6.3.2, Fig. 6.10. This architecture can be reconfigured to have a unitary gain, $G_{rect} = 1$, or to duplicate the input amplitude, $G_{rect} = 2$.

As deduced in Sect. 6.3.2, the rectifier input resistance can be approximated as

$$(6.3) \Rightarrow R_{rect} = \frac{R_L \cdot \eta_{rect}}{2G_{rect}^2}.$$

Therefore, an increase in R_L can be countered by an increase in the rectifier gain, to maintain R_{rect} closer to the optimum value.

The LTspice schematic with unitary gain ($G_{rect} = 1$) is the same presented in Fig. 8.12, while the one that is reconfigured with $G_{rect} = 2$ is shown in Fig. 8.13, supplementary material file Sec823.

In Table 8.14, we summarize the simulated and calculated values for $R_L = 500\,\Omega$, $R_L = 2500\,\Omega$, $G_{rect} = 1$, and $G_{rect} = 2$.

As can be seen from Table 8.14, the reconfigurable architecture is able to hold a high efficiency under loading variations, improving the efficiency from 19.5 % to 25.9 % with $R_L = 2500\,\Omega$ in simulations. The rectifier gain can be controlled on a closed-loop in different ways, for instance, by a Perturbation and Observation

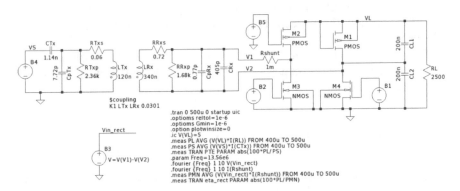

Fig. 8.13 LTspice schematic of the visual prosthesis with parallel matching network and reconfigurable rectifier at $G_{rect} = 2$. The LTspice schematic is available in the supplementary material, file Sec823

Table 8.14 Calculated and simulated main relevant values for $R_L = 500\,\Omega$, $R_L = 2500\,\Omega$, $G_{rect} = 1$, and $G_{rect} = 2$. The calculation script and LTspice schematics are available in the supplementary material, files Sec822, Sec822_parallel, and Sec823

	$G_{rect} = 1$				$G_{rect} = 2$			
	Fig. 8.12				Fig. 8.13			
	$R_L = 500\,\Omega$		$R_L = 2500\,\Omega$		$R_L = 500\,\Omega$		$R_L = 2500\,\Omega$	
	$P_L = 50\,mW$		$P_L = 10\,mW$		$P_L = 50\,mW$		$P_L = 10\,mW$	
	Calc.	Sim.	Calc.	Sim.	Calc.	Sim.	Calc.	Sim.
PTE (%)	26.4	25.0	20.5	19.5	13.7	13.8	27.3	25.9
R_{rect} (Ω)	250	280	1250	1283	63	81	313	342
	Closer to the MEP		Further from the MEP		Further from the MEP		Closer to the MEP	

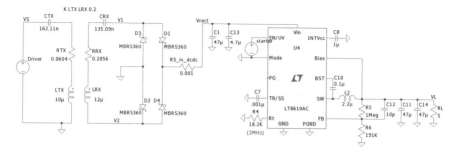

Fig. 8.14 LTspice schematic of WPT link designed for a smartphone charging pad. Supplementary material, file Sec830_spice

(P&O) method applied periodically or each time the load power consumption changes. Maintaining a high efficiency is desired to extend the battery life, which, in this case, is part of the external components of the visual prosthesis.

8.3 Smartphones

Most modern smartphones come with built-in wireless charging capability. One of the leading standards, supported by major smartphone manufacturers, is Qi. This standard was developed by the Wireless Power Consortium. Currently, the Qi standard is able to provide up to 15 W [12]. However, it is expected to be extended to laptop and tablet computers (30 to 60 watts). The operation frequency is in 87 to 205 kHz range [12]. Generally, the mobile device is laid on top of a charging pad; thus, the Tx to Rx distance is in the millimeter range.

Smartphones are charged at a DC voltage of 5 V at a power level that is around 5 W. Therefore, in this example, the load can be approximated as $5 \, \Omega = \frac{(5 \, V)^2}{5 \, W}$.

There are commercially available wireless charging coils, many of them characterized at 125 kHz, which is within the Qi band.

In this section, we provide an LTspice model of a WPT link which is reasonable for a smartphone charging pad, shown in Fig. 8.14. The operating frequency was set to 125 kHz, and the parameters of the coil are in the range of the commercially available coils that are designed and characterized at this frequency. In this example, to regulate the load voltage at $V_L = 5 \, V$, the LT8610AC synchronous step-down regulator was used in the Rx-circuit. Therefore, the output voltage regulation is implemented by a (D)-type feedback (see Fig. 7.1).

The Rx matching network of this example can be designed in the same manner as in Sect. 8.2.2 for the visual prosthesis. However, in this case, it is the series matching network the one that set a $Re\{Z_{MN}\}$ closer to the optimum value, $Re\{Z_{MN_{opt-\eta}}\}$, achieving higher PTE than the parallel topology.

In Sect. 7.5, we addressed the MEP tracking in links with the (D)-type output voltage regulation. It was shown that the Tx voltage, V_S, can be used to adjust the $Re\{Z_{MN}\}$. However, it was shown that the MEP, $Re\{Z_{MN}\} = Re\{Z_{MN_{opt-\eta}}\}$, cannot be achieved in all systems by just adjusting V_S. In Table 7.7, it was summarized which systems are able to achieve the MEP by adjusting V_S. Particularly, the system presented so far in this example (Fig. 8.14) which has a voltage source resonant Tx $\left(\frac{1}{\omega C_{Tx}} = \omega L_{Tx}\right)$, an even number of coils in the link (2-coil), and a series Rx matching network, is in row I of Table 7.7. Therefore, based on Table 7.7, the WPT link shown in Fig. 8.14 is not able to achieve the MEP by adjusting V_S. To solve this, as in the example presented at the end of Sect. 7.5, the Tx can be detuned. By detuning the Tx in this case, we can move from row I to row III in Table 7.7 and attain the MEP.

In Fig. 8.15, we plot the simulated PTE of this WPT link as a function of the Tx voltage source, V_S, with a tuned (as shown in Fig. 8.14) and a detuned Tx (increasing C_{Tx} from 162 nF to 192 nF). This detune can be justified using the analysis developed in Chap. 7, especially from Fig. 7.19. Figure 8.16 is similar to Fig. 7.19 but evaluated for this case. As can be seen from Fig. 8.16, the $Re\{Z_{MN}\}$ range with a resonant Tx ($C_{Tx} = \frac{1}{\omega^2 L_{Tx}} = 162$ nF, $X_S = 0$) does not include the MEP. However, the case with a non-resonant Tx ($C_{Tx} = 192$ nF, $X_S = 1.2\ \Omega$) does include the MEP.

As shown in Fig. 8.15, taking into account the analysis presented in Chap. 7, we were able to increase the PTE of the system presented in Fig. 8.14 from 40 % to 72 %. As a side effect, a higher voltage is required in the Tx as it sees a higher impedance.

To summarize, here we provided an example system which is designed for a smartphone charging pad by showing how the system efficiency can be strongly improved by applying the analysis developed in Chap. 7. The LTspice schematic is provided to allow the reader to see the result of the theoretical deduction in the actual system simulation and observe the impact of various circuit parameters.

8.4 Electric Vehicles

There are long-standing precedents for WPT for Electric Vehicles (EV), going as far back as patent 527.857 from 1894 [13], which describes a contactless power transfer method for electric railways.

In the last decades, the EV market has grown significantly fueled by the increased focus on renewable energy and maturation of the EV industry. EVs can either be almost constantly receiving power while moving from an off-vehicle source, such as in electric trains, or powered from a self-contained battery which is charged periodically as in electric cars. In both cases, WPT has proved to be a feasible and convenient method to provide power to the EV.

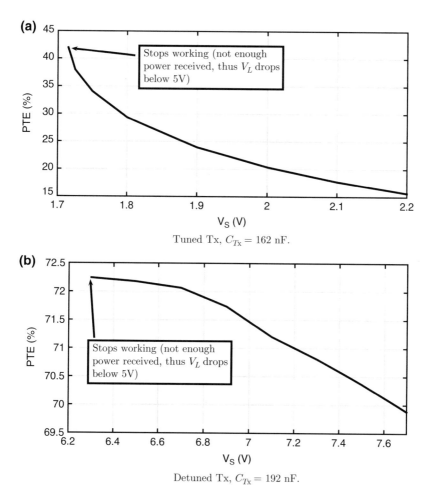

Fig. 8.15 Simulated PTE for the WPT example oriented to a smartphone charging pad, with a tuned Tx (**a**) and detuned Tx (**b**). Supplementary material, file Sec830_spice

In this application, a large amount of power is required by the load (the EV). Therefore, these systems involve larger currents than the previous examples and generate a relatively intense electromagnetic interference, which should comply with the safety regulations and human health concerns that were discussed in Sect. 3.3. Most designers use ferrite shields to reduce the magnetic fields surrounding the EV charging area, which if designed properly can even improve the PTE.

A detailed study of WPT for EV can be found in references such as [14]. In this section, we summarize an example link designed in [15] for a golf cart. This system was designed to provide 1 kW at 150 V DC, i.e., $R_L = \frac{(150 \text{ V})^2}{1 \text{ kW}} = 22.5 \ \Omega$. The carrier frequency was selected at 20 kHz, and the target distance between coils was $D_{Tx\text{-}Rx} = 15.6$ cm. In this system, identical coils were used for the Tx and Rx. The

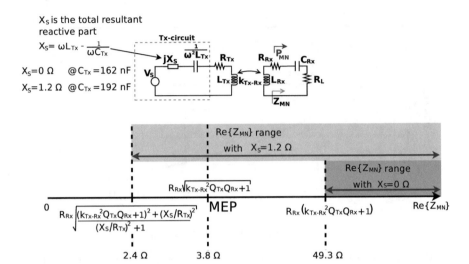

Fig. 8.16 Similar to Fig. 7.19 but evaluated for the example presented in this section

Fig. 8.17 Tx and Rx coil used in the WPT link example for EV. Designed in [15]

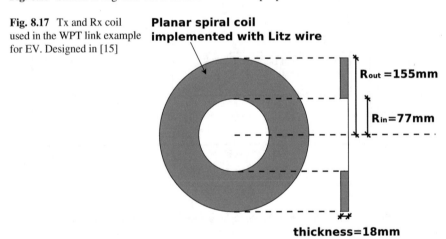

coils have a planar spiral shape and were implemented using Litz wire (see Fig. 3.1 and Table 3.1 for wire types and coil shape descriptions, respectively). The coils' design was based on a combination of theoretical formulas and simulation tools, and they were constrained to a maximum radius of 30 cm. The design procedure described in [15] for this system leads to the coil dimensions depicted in Fig. 8.17.

This example has an output power that is more than two orders of magnitude greater than the smartphone example, more than four orders of magnitude greater than the visual prosthesis example, and more than six orders of magnitude greater than the RFID example. However, all the analyses presented throughout the book to estimate efficiency; to determine the most appropriate Rx matching network, e.g.,

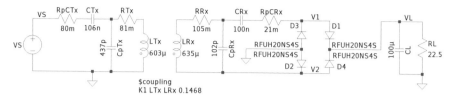

Fig. 8.18 LTspice schematic of WPT link designed in [15] for a golf cart

series or parallel topology; to track the OOP; or to regulate the output voltage can be used in all the examples presented in this chapter.

For this example, we provide the LTspice schematic which implements the previously described WPT link for a golf cart (Fig. 8.18), together with a calculation script that uses the expressions deduced in the book and were used in all the previously presented examples to estimate PTE (supplementary material, files Sec840_spice and Sec840). The calculated link efficiency is $\eta_{Link} = 97.5\%$ while in simulation $\eta_{Link} = 97.1\%$. The full-wave rectifier has, in simulation, a $\eta_{rect} = 98.4\%$; thus, the total PTE of the link is 95.6%.

References

1. K. Finkenzeller, *RFID Handbook: Fundamentals and Applications in Contactless Smart Cards, Radio Frequency Identification and Near-Field Communication* (Wiley, 2010)
2. ICAR, International agreement of recording practices, June 2008. www.icar.org
3. ISO, ISO 11785 radio frequency identification of animals – technical concept. Standard, International Organization for Standardization (1996)
4. P. Pérez-Nicoli, A. Rodríguez-Esteva, F. Silveira, Bidirectional analysis and design of RFID using an additional resonant coil to enhance read range. IEEE Trans. Microw. Theory Techn. **64**(7):2357–2367 (2016). ISSN 0018-9480. https://doi.org/10.1109/TMTT.2016.2573275
5. M. Ghovanloo, K. Najafi, A wireless implantable multichannel microstimulating system-on-a-chip with modular architecture. IEEE Trans. Neural Syst. Rehabil. Eng. **15**(3), 449–457 (2007)
6. J.D. Weiland, M.S. Humayun, Visual Prosthesis. Proc. IEEE **96**(7), 1076–1084 (2008)
7. U.-M. Jow, M. Ghovanloo, Design and optimization of printed spiral coils for efficient transcutaneous inductive power transmission. IEEE Trans. Biomed. Circuits Syst. **1**(3), 193–202 (2007)
8. U.-M. Jow, M. Ghovanloo, Modeling and optimization of printed spiral coils in air, saline, and muscle tissue environments. IEEE Trans. Biomed. Circuits Syst. **3**(5), 339–347 (2009)
9. K. Agarwal, R. Jegadeesan, Y.-X. Guo, N.V. Thakor, Wireless power transfer strategies for implantable bioelectronics. IEEE Rev. Biomed. Eng. **10**, 136–161 (2017)
10. H. Naganuma, K. Kiyoyama, T. Tanaka, A 37× 37 pixels artificial retina chip with edge enhancement function for 3-D stacked fully implantable retinal prosthesis, in *IEEE Biomedical Circuits and Systems Conference* (IEEE, 2012), pp. 212–215
11. P.M. Lewis, J.V. Rosenfeld, Electrical stimulation of the brain and the development of cortical visual prostheses: an historical perspective. Brain Res. **1630**, 208–224 (2016)
12. Wireless-Power-Consortium, Qi standard, power class 0 specification (2017). https://www.wirelesspowerconsortium.com/

13. H. Maurice, L. Maurice, Transformer system for electric railways. Oct 1894. US Patent 527,857
14. C.T. Rim, C. Mi, *Wireless Power Transfer for Electric Vehicles and Mobile Devices* (Wiley, 2017)
15. H. Kim, C. Song, D.H. Kim, D.H. Jung, I.M. Kim, Y.I. Kim, J. Kim, S. Ahn, J. Kim, Coil design and measurements of automotive magnetic resonant wireless charging system for high-efficiency and low magnetic field leakage. IEEE Trans. Microw. Theory Techn. **64**(2), 383–400 (2016). ISSN 0018-9480. https://doi.org/10.1109/TMTT.2015.2513394

Index

Printed in the United States
by Baker & Taylor Publisher Services